AF363302

Miniaturized Silicon Photodetectors

Miniaturized Silicon Photodetectors: New Perspectives and Applications

Editor

Maurizio Casalino

MDPI • Basel • Beijing • Wuhan • Barcelona • Belgrade • Manchester • Tokyo • Cluj • Tianjin

Editor
Maurizio Casalino
Institute of Applied Sciences and
Intelligent Systems
Italy

Editorial Office
MDPI
St. Alban-Anlage 66
4052 Basel, Switzerland

This is a reprint of articles from the Special Issue published online in the open access journal *Micromachines* (ISSN 2072-666X) (available at: https://www.mdpi.com/journal/micromachines/special_issues/Silicon_Photodetectors).

For citation purposes, cite each article independently as indicated on the article page online and as indicated below:

LastName, A.A.; LastName, B.B.; LastName, C.C. Article Title. *Journal Name* **Year**, *Volume Number*, Page Range.

ISBN 978-3-0365-0044-7 (Hbk)
ISBN 978-3-0365-0045-4 (PDF)

Cover image courtesy of Maurizio Casalino.

Contents

About the Editor

Maurizio Casalino (Dr.) graduated summa cum laude in electronic engineering from the University of Naples "Federico II" and he received his Ph.D. degree in engineering electronic from the University "Mediterranea" of Reggio Calabria in the 2008.

In the 2010, he became a researcher at the Institute of Applied Science and Intelligent Systems "Eduardo Caianiello" of the National Council of Research (ISASI-CNR) in Naples (Italy).

He is the author of more than 90 scientific articles published in peer-reviewed journals, proceedings of congress, book chapters, and review articles. His research interests include the design, fabrication, and characterization of photonic devices. Dr. Casalino has served as a reviewer for many international scientific journals, including *Nanoscale*, *Small*, *ACS Nano*, and *Nature Communications*.

He is a member of the Italian Society of Optics and Photonics (SIOF), member of the Italian Society of Electronics (SIE), and member of the Institute of Electrical and Electronics Engineers (IEEE).

Editorial

Editorial for the Special Issue on Miniaturized Silicon Photodetectors: New Perspectives and Applications

Maurizio Casalino

Institute of Applied Science and Intelligent Systems "Eduardo Caianiello" (CNR), Via P. Castellino n. 141, 80131 Naples, Italy; maurizio.casalino@na.isasi.cnr.it

Received: 9 November 2020; Accepted: 13 November 2020; Published: 17 November 2020

Silicon (Si) technologies provide an excellent platform for monolithically integrating both photonic [1] and microelectronic [2] functionalities in the same substrate. In the last few years, a variety of passive and active Si photonic and optoelectronic devices have been reported, in particular in the field of photodetection [3–5] where new effects and structures have been proposed. Si photodetectors (PDs) at visible wavelengths are a commercial reality, but, unfortunately, they cannot be employed in the infrared (IR) range due the Si transparence over 1100 nm. Historically, the use of germanium (Ge) grown on Si has allowed for the realization of Si-based PDs up to 1550 nm [6]. In recent years, impressive progresses have been achieved by extending the operation of Si PDs to the infrared regime and, recently, thanks to the investigation of new smart structures and effects, the all-Si approach has demonstrated potentialities leading to performances close to those of the well-established germanium (Ge) technology. Moreover, the possibility to integrate new emerging 3D and 2D materials with Si, together with the capability of manufacturing devices at the nanometric scale, has led to the development of new devices with unexpected performance.

There are eight papers published in this Special Issue, six original works and two review articles. The spectral range covered by these works goes from the ultraviolet (UV) to the infrared (IR) regime. Among the original works, four of them are focused on the investigation of novel Si-based PDs while the remaining two on the characterization of materials that could be employed in Si technology. The four papers proposing original devices are based on Schottky, metal–semiconductor–metal (MSM) and P/N structures. The proposed Schottky photodetectors take advantage of the integration of graphene on both crystalline silicon (c-Si) and polycrystalline silicon (poly-Si) substrates while the MSM PDs are based on germanium–tin (GeSn) layers. Both structures (Schottky and MSM) have shown the capability to detect near infrared (NIR) wavelengths. On the other hand, a complementary metal–oxide–semiconductor (CMOS) single photon avalanche P/N diode has been investigated for detecting light in the visible regime. Concerning the two papers dealing with the material characterization, they investigate layers of 4H-SiC and ZnO to be integrated on Si for detecting UV and visible regime. Finally, two review articles are focused on the development of CMOS-compatible microbolometers and integrated PDs based on group IV and colloidal semiconductors.

In particular, Tsai et al. [7] proposed a graphene/poly-Si PD to monolithically integrate with the electronic circuitry constituting the active pixel of a CMOS image sensor. This work is interesting mainly for two reasons. First, although graphene/crystalline Si PDs have been frequently reported in the literature [8], the investigation of graphene/polycrystalline junctions is much less frequently discussed. Second, the use of polycrystalline silicon as semiconductor, instead of crystalline silicon, makes the photodetector able to be directly integrated on top of the gate oxide of a conventional metal–oxide–semiconductor field effect transistor (MOSFET), enabling the realization of a compact active pixel to be employed in CMOS image sensors. The authors name this new proposed structure as: photodiode–oxide–semiconductor field effect transistor (PDOSFET). If the device described in the work of Tsai et al. should operate in the visible range, M. Casalino theoretically investigates the possibility of employing hybrid graphene/c-Si Schottky diodes to detect NIR wavelengths [9].

In this work, the absorption mechanism is based on the internal photoemission effect: graphene first absorbs the incoming radiation and then it transfers the photoexcited carriers into Si where they are collected. In addition, this work suggests integrating the graphene layer in the middle of a silicon-based Fabry–Pérot microcavity constituted by an amorphous hydrogenated silicon/graphene/crystalline silicon three-layer system surrounded by two high reflectivity mirrors. The author shows that the enhancement of the optical field inside the cavity allows a significant increase in graphene optical absorption and, consequently, in device efficiency. Theoretical results show responsivity of 0.24 A/W, bandwidth in GHz regime, noise equivalent power of 0.6 nW/cm$\sqrt{}$Hz. MSM PDs have been proposed by Ghosh et al. taking advantage of GeSn layers integrated on Ge-buffered Si substrates for short-wave infrared (SWIR) applications [10]. Indeed, GeSn shows a significant absorption along the entire telecommunication bands unlike germanium (Ge), whose optical absorption falls drastically beyond 1550 nm. GeSn MSM PDs have been both electrically and electro-optically characterized. The I–V electrical characteristic shows the classical MSM behavior while the spectral responsivity measurements show a broadband optical absorption extending over 1800 nm. The reported responsivity increases by increasing the bias voltage and at 7 V maximum values of about 100, 70 and 10 mA/W at 1200, 1500 and 1800 nm, have been reported, respectively. The paper of Goll et al. investigates the discharge mechanism of single-photon avalanche diodes (SPADs) designed in 0.35 μm CMOS technology [2]. Indeed, after the avalanche has been triggered, the SPAD cathode–anode voltage reaches the breakdown voltage with a time that has been measured by the authors. Based on the cathode capacitance measurements, the avalanche current through each SPAD was evaluated too. Measurements on the cathode voltage transient of SPADs based on a 12 μm-thick p⁻ epi-layer (named type A) with various diameters were investigated. Results show fall times of 3.45 ns for 200 μm diameter SPAD and an excess bias (voltage difference between diode work reverse voltage and the breakdown voltage) of 4.26 V, as well as fall times of 10.2 ns for 50 μm diameter SPAD and an excess bias of 4.2 V. On the other hand, SPADs with different diameters were implemented in the high-volume (HV) line of the same CMOS process (named type B) showing fall times of 2 ns for 98.2 μm diameter SPAD and 5.9 V excess bias, as well as 8 ns for 48.2 μm diameter SPAD and 5.4 V excess bias.

Moving our attention onto the characterization of materials to be employed for the realization of Si-based photodetectors, J. Li et al. investigate how different chemical vapor deposition (CVD) growth conditions impact the defect density of 4H-SiC epilayers and the performance of Nickel(Ni)/4H-SiC Shottky PDs [11]. In this work, particular attention was paid to triangular defects (TDs) and deep level defects, $Z_{1/2}$, showing that, while the C/Si ratio strongly impacts the formation of TDs, no correlation with the deep level defect, $Z_{1/2}$, can be confirmed. This work shows that, by adjusting the C/Si ratio, the quality of the 4H-SiC epilayer can be improved and the performance of the Ni/4H-SiC Schottky detector increased. In their work, G. Li et al. evaluate the impact ionization coefficient of electrons in a ZnO layer along the (001) direction [12]. In order to do it, the authors have investigated p-Si/i-ZnO/n-AZO structures illuminated by a 532 nm laser diode where the electron avalanche multiplication is triggered in the i-ZnO layer whose thickness was varied from 250 to 750 nm. These insights could be very useful for the realization of high-performance ultraviolet (UV) avalanche photodiodes).

Finally, this Special Issue includes also two review articles: Dardano and Ferrara have reported recent advances in the field of photodetection based on group IV materials with particular reference to silicon [13]. In recent years, the challenge to make silicon usable at NIR wavelengths has attracted much interest and many absorption mechanisms have been both proposed and investigated. The authors show as the mid-bandgap absorption (MBA), i.e., the infrared absorption obtained by voluntarily introducing defects in the Si bandgap, combined with high Q-factor cavity structures (ring resonators), has emerged as a viable solution for the realization of devices whose performance are comparable with the well-established Ge technology. Then, in their work, the authors reviewed PDs based on different materials, such as graphene, Ge and carbon nanotubes (CNTs). Moreover, an overview on PDs based on colloidal semiconductors, representing the frontier of future research, has been presented too. Finally, Yu et al. have reviewed the CMOS microbolometer technology for long-wave infrared

(LWIR) imaging applications at a low cost [14]. This technology is based on a standard CMOS process combined with a simple post-CMOS micro-electro-mechanical system (MEMS) process. In their work, the authors show that the performance of the reported CMOS-compatible microbolometers has started to compare favorably with the state of the art. This paper reviews not only the recent advances of the CMOS-compatible microbolometers but also the aspects of the pixel structure and of the read-out integrated circuitry.

I would like to take this opportunity to thank all the authors for submitting their papers to this Special Issue. I would also like to thank all the reviewers for dedicating their time and helping to improve the quality of the submitted papers. Finally, I would like to warmly thank Ms. Aria Zeng for her constant support in preparing this Special Issue.

Funding: This research received no external funding.

Conflicts of Interest: The author declares no conflict of interest.

References

1. Managò, S.; Zito, G.; Rogato, A.; Casalino, M.; Esposito, E.; De Luca, A.C.; De Tommasi, E. Bioderived Three-Dimensional Hierarchical Nanostructures as Efficient Surface-Enhanced Raman Scattering Substrates for Cell Membrane Probing. *ACS Appl. Mater. Interfaces* **2018**, *10*, 12406–12416. [CrossRef] [PubMed]

2. Goll, B.; Steindl, B.; Zimmermann, H. Avalanche Transients of Thick 0.35 μm CMOS Single-Photon Avalanche Diodes. *Micromachines* **2020**, *11*, 869. [CrossRef] [PubMed]

3. Casalino, M. Design of Resonant Cavity-Enhanced Schottky Graphene/Silicon Photodetectors at 1550 nm. *J. Ligthwave Technol.* **2018**, *36*, 1766–1774. [CrossRef]

4. Zhu, S.; Chu, H.S.; Lo, G.Q.; Bai, P.; Kwong, D.L. Waveguide-integrated near-infrared detector with self-assembled metal silicide nanoparticles embedded in a silicon p-n junction. *Appl. Physics Lett.* **2012**, *100*, 061109. [CrossRef]

5. Casalino, M.; Russo, R.; Russo, C.; Ciajolo, A.; Di Gennaro, E.; Iodice, M. Free-Space Schottky Graphene/Silicon Photodetectors operating at 2 μm. *ACS Photonics* **2018**, *5*, 4577–4585. [CrossRef]

6. Koester, S.J.; Schaub, J.D.; Dehlinger, G.; Chu, J.O. Germanium-on-SOI infrared detectors for integrated photonic applications. *IEEE J. Selected Topics Quant. Electron.* **2006**, *12*, 1489–1502. [CrossRef]

7. Tsai, Y.-Y.; Kuo, C.-Y.; Li, B.-C.; Chiu, P.-W.; Hsu, K.Y.J. A Graphene/Polycrystalline Silicon Photodiode and Its Integration in a Photodiode–Oxide–Semiconductor Field Effect Transistor. *Micromachines* **2020**, *11*, 596. [CrossRef] [PubMed]

8. Amirmazlaghani, M.; Raissi, F.; Habibpour, O.; Vukusic, J.; Stake, J. Graphene-Si Schottky IR detector. *IEEE J. Quantum Electron.* **2013**, *49*, 589. [CrossRef]

9. Casalino, M. Theoretical Investigation of Near-Infrared Fabry–Pérot Microcavity Graphene/Silicon Schottky Photodetectors Based on Double Silicon on Insulator Substrates. *Micromachines* **2020**, *11*, 708. [CrossRef] [PubMed]

10. Ghosh, S.; Lin, K.-C.; Tsai, C.-H.; Kumar, H.; Chen, Q.; Zhang, L.; Son, B.; Tan, C.S.; Kim, M.; Mukhopadhyay, B.; et al. Metal-Semiconductor-Metal GeSn Photodetectors on Silicon for Short-Wave Infrared Applications. *Micromachines* **2020**, *11*, 795. [CrossRef] [PubMed]

11. Li, J.; Meng, C.; Yu, L.; Li, Y.; Yan, F.; Han, P.; Ji, H. Effect of Various Defects on 4H-SiC Schottky Diode Performance and Its Relation to Epitaxial Growth Conditions. *Micromachines* **2020**, *11*, 609. [CrossRef] [PubMed]

12. Li, G.; Zhao, X.; Jia, X.; Li, S.; He, Y. Characterization of Impact Ionization Coefficient of ZnO Based on a p-Si/i-ZnO/n-AZO Avalanche Photodiode. *Micromachines* **2020**, *11*, 740. [CrossRef] [PubMed]

13. Dardano, P.; Ferrara, M.A. Integrated Photodetectors Based on Group IV and Colloidal Semiconductors: Current State of Affairs. *Micromachines* **2020**, *11*, 842. [CrossRef] [PubMed]

14. Yu, L.; Guo, Y.; Zhu, H.; Luo, M.; Han, P.; Ji, H. Low-Cost Microbolometer Type Infrared Detectors. *Micromachines* **2020**, *11*, 800. [CrossRef] [PubMed]

Publisher's Note: MDPI stays neutral with regard to jurisdictional claims in published maps and institutional affiliations.

Article

Avalanche Transients of Thick 0.35 μm CMOS Single-Photon Avalanche Diodes

Bernhard Goll *, Bernhard Steindl and Horst Zimmermann

Institute of Electrodynamics, Microwave and Circuit Engineering, TU Wien, Gusshausstrasse 25/E354-02, A-1040 Wien, Austria; bernhard.steindl@tuwien.ac.at (B.S.); horst.zimmermann@tuwien.ac.at (H.Z.)
* Correspondence: bernhard.goll@tuwien.ac.at

Received: 1 September 2020; Accepted: 17 September 2020; Published: 19 September 2020

Abstract: Two types of single-photon avalanche diodes (SPADs) with different diameters are investigated regarding their avalanche behavior. SPAD type A was designed in standard 0.35-μm complementary metal-oxide-semiconductor (CMOS) including a 12-μm thick p$^-$ epi-layer with diameters of 50, 100, 200, and 400 μm; and type B was implemented in the high-voltage (HV) line of this process with diameters of 48.2 and 98.2 μm. Each SPAD is wire-bonded to a 0.35-μm CMOS clocked gating chip, which controls charge up to a maximum 6.6-V excess bias, active, and quench phase as well as readout during one clock period. Measurements of the cathode voltage after photon hits at SPAD type A resulted in fall times (80 to 20%) of 10.2 ns for the 50-μm diameter SPAD for an excess bias of 4.2 V and 3.45 ns for the 200-μm diameter device for an excess bias of 4.26 V. For type B, fall times of 8 ns for 48.2-μm diameter and 5.4-V excess bias as well as 2 ns for 98.2-μm diameter and 5.9-V excess bias were determined. In measuring the whole capacitance at the cathode of the SPAD with gating chip connected, the avalanche currents through the detector were calculated. This resulted in peak avalanche currents of, e.g., 1.19 mA for the 100-μm SPAD type A and 1.64 mA for the 98.2-μm SPAD type B for an excess bias of 5 and 4.9 V, respectively.

Keywords: single-photon avalanche diode (SPAD); gating; avalanche transients; 3.3 V/0.35 μm complementary metal-oxide-semiconductor (CMOS)

---

1. Introduction

Avalanche photodiodes (APDs) operated with a reverse voltage larger than the breakdown voltage (Geiger mode) are usually capable of detecting single photons and are termed single-photon-avalanche-diodes (SPADs). A photon absorption generates an electron-hole pair, and in combination with the high-electric field in a multiplication zone, a large avalanche with charge carriers might be triggered due to impact ionization. This causes a current, which discharges the SPAD until its cathode-anode voltage reaches the breakdown voltage, where the avalanche is quenched. For a further detection of a photon, the SPAD has to be recharged again.

With reference to [1], the avalanche build-up (time between electron-hole pair generation due to a photon hit and reaching maximum avalanche charge) action of a SPAD consists at first in a local charge multiplication, a local voltage drop to breakdown level and then a spreading to lateral directions. After build-up, final quenching happens. The spreading might take place with the help of charge carriers, which move towards side directions or create additional secondary photons when some avalanche carriers recombine. The simplest equivalent circuit to model an avalanche action is the capacitance of the reverse biased avalanche diode in parallel with a resistor [1,2]. When a photon enters a SPAD, the photon detection probability (PDP) describes the chance that a self-sustaining avalanche is triggered. Even in the absence of photons, dark counts occur in a SPAD, which are uncorrelated avalanches due to thermal-/trap-assisted carrier generation or tunneling. They are characterized by a mean dark count rate (DCR). Afterpulses, on the other hand, are avalanches, which are correlated to a previous avalanche. There are many reasons for afterpulses, e.g., the release of a carrier by a deep-level trap, which has been filled during a previous current flow, or, e.g., the diffusion of secondary charge carriers into the high-field zone of the SPAD, which were generated by photons originating from recombination during avalanche current flow. The probability for appearance of an afterpulse is described with afterpulsing probability (APP). It becomes lower the more time elapses after a previous avalanche. The voltage difference between how much the diode's reverse-voltage is higher than the breakdown voltage is one of the main parameters in operating a SPAD and is called excess bias. Typically, DCR, APP, and PDP increase when raising the excess bias [3–6]. After an avalanche has happened in a SPAD, it needs a dead time, which is controlled by surrounding circuitry, until it is recharged again and ready for a new photon detection. The APP strongly depends on the dead time. If the dead time is longer, the APP will decrease.

SPADs are important in applications like photon detectors for quantum communications [7,8] and quantum random number generators [9–11]. Recently, many multi-pixel image sensors with SPADs were published [12–17], some for 3D imaging. SPAD arrays have potential for highly sensitive optical data receivers [18–21]. Typically, the bit error rate (BER) of SPAD receivers suffers from DCR and APP. Forward error correction might be a solution to solve this problem [22]. A BER of 2×10^{-3} is sufficient to use a concatenated Reed–Solomon code super-forward-error-correction (FEC) scheme to get a BER better than 10^{-9} with 6.69% redundancy (ITU-T G.975.1). In [23], a 64×64 SPAD array in 130 nm complementary metal-oxide-semiconductor (CMOS) was capable to receive a 500 Mb/s 4-PAM optical signal with −46.1 dBm sensitivity for a bit error rate (BER) of 2×10^{-3} when using equalization. In [24], five subsequent time slots (one period of a 250 MHz clock), where each slot consists of the information of whether one or no photon was detected from a gating circuit in a 3.3-V/0.35-μm CMOS technology with one SPAD are fed into a shift register. Hence it could be decided whether a bit has been received or not with the knowledge of how many detections happened in a row of five time slots. This fully integrated optical receiver achieved a data rate of 50 Mb/s in non-return-to-zero (NRZ) with a sensitivity of −57 dBm (BER = 2×10^{-3}).

This paper presents measurement results of the cathode voltage drop of two types of SPADs with different diameters when an avalanche occurs. SPAD type A was fabricated in standard 0.35-μm CMOS with 12-μm thick p⁻ epi-layer with active diameters of 50, 100, 200, and 400 μm and type B was implemented in the high-voltage (HV) line of this process with active diameters of 48.2 and 98.2 μm. With the knowledge of the measured capacitance at the cathode node, the avalanche currents through each SPAD were determined. The voltage transient response with a measurement of its 80 to 20% fall time of SPAD type A with 50 μm diameter was already published in [25]. Each SPAD is wire-bonded to a clocked gating chip, which controls charge up to maximum 6.6-V excess bias, active, and quench phase as well as readout during one clock period. This cascaded gating chip was designed in a standard 3.3-V/0.35-μm CMOS technology.

2. Single-Photon Avalanche Diodes (SPADs)

In Figure 1, cross sections of SPAD type A (see Figure 1a) and SPAD type B (see Figure 1b) are depicted. Both diodes were fabricated in a 3.3-V/0.35-μm CMOS technology. Type A had a ≈12-μm thick low-doped p⁻ epi layer with a doping concentration of $\approx 2 \times 10^{13}$/cm^3 and type B had the epitaxial layer of the high-voltage process version, which was doped with $\approx 10^{15}$/cm^3 [26]. Each type of SPAD could be integrated together with additional circuitry on one chip each [24] when the electronic part was isolated from the substrate with the help of deep n-wells. SPAD type A used the standard CMOS process line and consisted of a n⁺⁺ cathode with a p-well below to form a high-field multiplication zone. A thick p⁻ epi layer acted as an absorption zone with a high-enough electric field for a high drift velocity of photo-generated charge carriers. Therefore, the maximum sensitivity of the SPAD was located in the visible red and was very near infrared region of the spectrum of the light. An n-well around the multiplication zone prevented the SPAD from edge breakdown. SPAD type A was fabricated in diameters of 50, 100, 200, and 400 μm. Typical values for its DCR were, e.g., $\approx 10^4$ counts per second (CpS) and for APP, e.g., 0.2% for a diameter of 50 μm at 20 °C, an excess bias of 3 V and a dead time of 9.5 ns [5]. The value of the PDP amounted to 21.5% for a wavelength of 635 nm [24], where the excess bias typically was near below ≈ 3 V for best BER. For a diameter of 100 μm, the DCR amounted to 1.89×10^4 CpS and 3.08×10^4 CpS for an excess bias of 3.3 and 6.6 V, respectively, at a temperature of 25 °C [27]. The APP was 0.7 and 4.8% for an excess bias of 3.3 and 6.6 V, respectively, and a dead time of 9.5 ns. For an excess bias of 6.6 V and wavelengths of 635 and 850 nm, PDPs of 35.1 and 22%, respectively, were achieved.

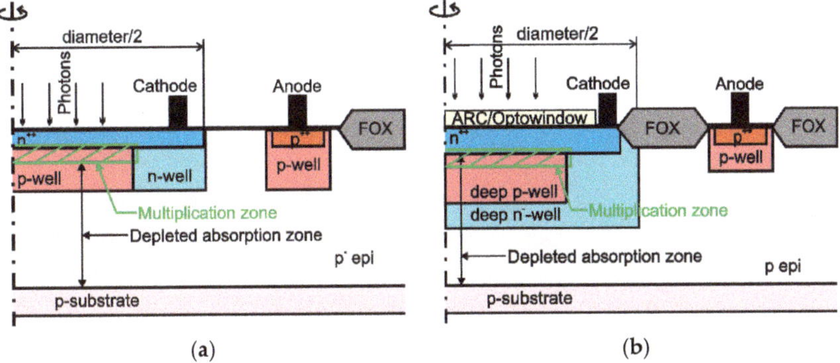

Figure 1. Cross sections (not to scale) of (**a**) single-photon avalanche diode (SPAD) type A in 0.35 μm standard CMOS with p⁻ epi-layer and (**b**) SPAD type B in 0.35 μm high-voltage (HV) complementary metal-oxide-semiconductor (CMOS) with standard p epi-layer, deep p-well, and deep n⁻-well for high-voltage applications (FOX = field oxide).

SPAD type B was designed in the high-voltage (HV) line of this 0.35-μm CMOS technology with the option of an oxide opening (opto window) with anti-reflection coating (ARC) above the diode (see Figure 1b), which was a nitride layer, where the thickness was optimized for no reflection at visible red light. The cathode consisted of a highly doped n⁺⁺ region, which was slightly thicker than the cathode of type A. The high-field multiplication zone was located at the passage between cathode and a deep p-well. To assure that the electric field in the depleted absorption zone, which has the same function as for SPAD type A, was high enough for a high drift velocity of photo-generated charge carriers, a deep n⁻-well was added. SPAD type B was fabricated in diameters of 49.2 and 98.2 μm. At a temperature of 25 °C and for a diameter of 49.2 μm, the best measured DCR amounts to 2.88×10^4 CpS for samples in the middle of the wafer up to 14.04×10^4 CpS for samples from the border for an excess bias of 6.6 V. The APP was determined to be near 80% for samples in the middle and down to 10% for samples from the border of the wafer if a dead time of 5.8 ns was used [28]. The PDP was 37.4, 27.9, and 18.6% for wavelengths of the light of 780, 850, and 900 nm, respectively. For the SPAD type B with a diameter of 98.2 μm, a DCR of $\approx 5.5 \times 10^5$ CpS for an excess bias of 3.3 V at a temperature of 25 °C was measured [29]. The APP was $\approx 10\%$ for an excess bias of 3.3 V and 6 ns dead time. The measured PDP for an excess bias of 3.2 V was $\approx 21\%$ for 650 nm wavelength. For a wavelength of 800 nm and 6.6 V excess bias, a PDP of 35% was obtained.

A figure of merit (FoM) to compare the performance of different detectors is the noise-equivalent power (NEP) [28,30]. It is depicted in Equation (1), where h is the Planck constant, c is the speed of light in vacuum, and λ the wavelength of the used light.

$$NEP = \frac{hc}{\lambda} \frac{\sqrt{2DCR}}{PDP} \tag{1}$$

For SPAD type A with a diameter of 50 μm, this resulted to $NEP \approx 205.8$ aW$\sqrt{\text{Hz}}$ for 3-V excess bias and 635-nm wavelength. For SPAD type B with a diameter of 49.2 μm, a best $NEP \approx 163.4$ aW$\sqrt{\text{Hz}}$ was calculated for 6.6-V excess bias and 780-nm wavelength. The focus of this paper is the transient measurement of photon-triggered avalanche pulses. More information about PDP, DCR, and APP (also in dependence on excess bias) of SPADs type A and B is published in [5,28,29].

3. Gating Chip

For controlling the SPADs, a gating chip was designed in 0.35-μm CMOS technology with a nominal supply voltage of 3.3 V. In comparison to a quenching circuit, a gated SPAD has defined, mostly periodic time slots, where it is set to active and ready for photon detection. Once a photon has triggered an avalanche, the SPAD is conducting until the breakdown voltage is reached by the cathode-anode voltage or until the reset phase starts (if the photon was absorbed close to the end of the active phase) and quenches the detector below breakdown voltage. This can be, e.g., done with a clock signal, which defines that in one half of the clock period the SPAD is active and in the other half it is quenched. In the active time window, at most, one avalanche can occur. Therefore, to be able to detect more photons, the clock frequency has to be increased. In the case of a data receiver, this results in a larger clock frequency than the data rate [24].

Figure 2 shows the block diagram of the gating control chip. The cathode of the SPAD was bonded with gold wire with 25-μm diameter and 1-mm length to the node CAT. Hence the gating controller can pull the cathode potential to $\approx V_{SPAD} = 3.3$ V to set the SPAD active for photon detection in Geiger mode or to $\approx V_{SS} = -3.3$ V to quench the SPAD in the reset phase. The resulting cathode-anode voltage is V_{SPAD}-V_{An} for detection and V_{SS}-V_{An} for reset. For operation, the breakdown level of the SPAD should be located somewhere in between. A detailed schematic of the whole gating control chip including the transients are depicted in Figure 3. On-chip, a clock driver generated digital clocks for the circuit block SPAD control and for the switching transistors out of a sine wave with $\approx$600 mV amplitude, which was applied to pad CLKIN. With a bond wire, CLKIN was connected to a 50-Ω micro-strip line on the printed circuit board (PCB). Therefore, an on-chip 50-Ω resistor was added for termination without appreciable reflections. The on-chip clock driver generates a digital non-inverted and inverted clock, nodes CLK, and $\overline{\text{CLK}}$, where the logical voltage levels were ground node GND = 0 V for digital low and $V_{DDL} = 3.3$ V for digital high. Clock signal $\overline{\text{CLKD}}$ corresponded to $\overline{\text{CLK}}$, but was level shifted down by 3.3 V so that the logical voltage levels result to −3.3 and 0 V.

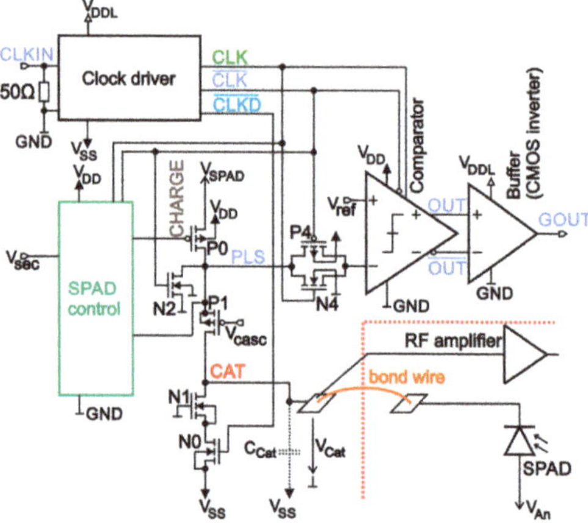

Figure 2. Block diagram of a clocked gating control chip, where a SPAD is connected with a bond wire. A radio frequency (RF) amplifier measures the cathode voltage via a RF probe needle on the pad at node cathode node (CAT).

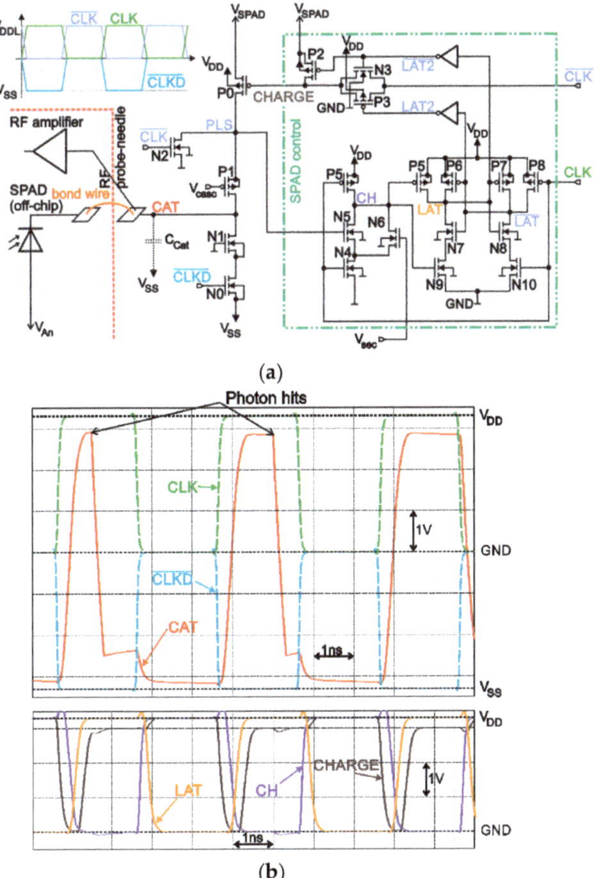

Figure 3. SPAD control in the gating control chip: (**a**) Schematics; (**b**) Simulated transients assuming a fast responding SPAD. A photon hit triggers an avalanche in the SPAD, which discharges node CAT until the cathode-anode voltage reaches its breakdown voltage and the avalanche is quenched.

The anode voltage of the external SPAD was applied via a separated pad, which was wire-bonded to the PCB. To charge up the cathode node (CAT) to $\approx V_{SPAD} = 3.3$ V, transistor P0 was turned on. To discharge node CAT to $\approx V_{SS} = -3.3$ V for quenching an avalanche in the SPAD, or to deactivate it, transistor N0 was switched on. All transistors are specified to work with a supply voltage of 3.3 V. To be able to switch node CAT between $\approx V_{SS} = -3.3$ V and $\approx V_{SPAD} = 3.3$ V, which resulted into a 6.6-V swing, cascode transistors P1 and N1 were added to protect all transistors against overvoltage. N-MOS transistors are typically smaller and faster with less parasitic capacitances than P-MOS transistors. Therefore, it was sufficient to connect the gate of N1 to GND, whereas on the gate of P1, a voltage of $V_{casc} = -1$ V was applied to faster charge-up node CAT and to reduce voltage peaks in the drain-source voltage of P0 and P1, which could exceed their critical voltages during switching action. On the other side, transistor N2 helped discharging node PLS due to similar reasons.

A clock period consisted of a reset phase (CLK = $\overline{\text{CLKD}}$ = GND = 0 V and $\overline{\text{CLK}}$ = V_{DDL} = 3.3 V), where the cathode-anode voltage of the SPAD was below the breakdown voltage, when node CAT was pulled down to $\approx V_{SS} = -3.3$ V, $PLS \approx 0$ V, and an active phase where CLK switches to $V_{DDL} = 3.3$ V ($\overline{\text{CLK}}$ = GND = 0 V and $\overline{\text{CLKD}}$ = V_{SS} = -3.3 V). In reset of the SPAD, the node voltages in the unit SPAD control were $\overline{\text{LAT}}$ = CH = V_{DD} = 3.3 V and LAT = 0 V, thus transmission gate P3/N3 was turned on, P2 was off, and node CHARGE = $\overline{\text{CLK}}$ = V_{DDL} = 3.3 V, hence transistor P0 was off. Transistor N0 was turned on.

In the active phase, transistor N0 is off. At first, a small fraction of the time duration in this phase was used to charge up the SPAD in pulling up node CAT and PLS to $\approx V_{SPAD}$ = 3.3 V with P0 turned on, because at the beginning, nodes $\overline{LAT}$ = CH = V_{DD} and LAT = 0 V, thus transmission gate P3/N3 was on, P2 off, and CHARGE = $\overline{\text{CLK}}$ = 0 V. Transistor P0 was charging node CAT and PLS until PLS reached a voltage level near V_{SPAD}. This charge-up was monitored at node PLS with transistor N5, which is turned on after its gate-source voltage raises above its threshold voltage. As a consequence, node CH is discharged with transistor N4 to $\approx$ 0 V, which forces the latch to flip to LAT = V_{DD} and $\overline{LAT}$ = 0 V, thus transmission gate P3/N3 is turned off and P2 is turned on, which charges up node CHARGE to V_{SPAD} to stop charging up node CAT and PLS with transistor P0. Because of parasitic capacitances and delay times of logic elements, the time lag between detection with N5 and turning off P0 was sufficiently long that nodes CAT and PLS can easily reach a voltage level very close to V_{SPAD}. Following this, the cathode-anode voltage was above the breakdown voltage; transistors P0, N0, and N2 were turned off; node CAT was charged up and floating; and the SPAD was ready for photon detection. The excess bias depended on the size of the anode voltage V_{AN} of the SPAD. For a distinct breakdown voltage V_{BD}, the excess bias V_{EB} can be calculated to $V_{EB} \approx V_{SPAD} - V_{AN} - V_{BD}$, where the breakdown level of the cathode voltage must be located between V_{SPAD} and V_{SS}, hence $V_{EB} \leq V_{SPAD} - V_{SS}$.

There could be the case that during charge up of node CAT, when P0 is turned on, an avalanche might occur in the SPAD, which would result especially for low ohmic diodes in a large current flow. Consequently, transistor N5 would never detect a finished charge up at node PLS and P0 stays turned on until the subsequent reset phase. To avoid such a large current flow during nearly the whole active phase of the SPAD, transistor N6 was added. With voltage V_{sec} at the gate of N5, node CH can be discharged independently from transistor N5 during an adjustable time duration. Thus, with adjusting V_{sec}, transistor N6 can be turned off or a discharging time for node CH can be set.

To read out whether an avalanche occurred or not, transmission gate N4/P4 was turned on during the active phase of the SPAD. The clocked comparator was in reset. In the subsequent reset phase of the SPAD, transmission gate N4/P4 was turned off and the voltage at node PLS was stored dynamically in the parasitic capacitance at the negative input-node of the comparator at the end of the active phase. A voltage drop indicated an avalanche. In the reset phase of the SPAD the comparator compared the stored voltage with a reference voltage V_{ref} to generate a digital decision at node OUT dependent on whether an avalanche occurred or not. With V_{ref}, the detection threshold for the voltage drop at PLS can be set. Finally, a 50-Ω driver was implemented, which consisted of a chain of inverters capable to drive an off-chip 50 Ω load, i.e., a fast oscilloscope.

Capacitance C_{Cat} in Figures 2 and 3a represents the overall node capacitance of node CAT including the cathode of the SPAD. Neglecting the bond wire in between due to its low inductance was justified. When an avalanche occurs during the active phase, transistors N0, P0 and N2 were off and only the current through the SPAD can discharge node CAT. Consequently, the avalanche current can be calculated out of the transient of the avalanche with $C_{Cat} \times dV_{Cat}/dt$ when measuring the whole node capacitance C_{Cat}, which included the capacitance of the SPAD due to its connection with a bond wire.

4. Measurement Setup, Results, and Discussion

Each SPAD was glued together with one gating chip on a printed circuit board (PCB) consisting of FR4 base material (FR = flame retardant). The cathode of the SPAD was bonded with a gold bond wire to node CAT of the gating chip. The anode of the SPAD was bonded to a DC line on the PCB, which provides the negative anode voltage V_{AN}. The sine wave to generate on-chip a digital clock signal was applied via a microstrip line and an SMA connector on the PCB to the gating chip. All supply and reference voltages were provided to the chip via connectors, block capacitances, and lines on the

PCB as well. The PCB itself was mounted on a copper block with temperature sensor and with a Peltier cooler below, which regulated the temperature on the PCB to 25 °C for measurements.

The transients of the cathode voltage at the pad of node CAT (see Figure 1) were measured with a high-speed radio frequency (RF) probe, Picoprobe Model 35 from GGB Industries, which had a frequency response from DC to 26 GHz, an operating range of −6 to 6 V, 10:1 signal attenuation, and 1.25-MΩ and 50-fF load at the input. The Model 35 Picoprobe was connected with a coaxial cable (K-system) to a Keysight MSOV204A Mixed Signal Oscilloscope with 20 GHz bandwidth and at maximum 80-GSa/s sampling rate. Unfortunately, the anode pad of the SPAD was situated nearby the cathode pad. Therefore the needle was placed on the pad of node CAT to avoid an accidental touch to the anode pad with its large negative voltage, and consequently, to damage of the expensive Picoprobe. The length of the bond wire between the cathode of the SPAD and node CAT of the gating chip was ≈2 mm, which corresponds to a series inductivity of ≈2 nH when using 1 nH/mm as a rule of thumb. A comparison of the transient at node CAT with the transient directly at the cathode pad of the SPAD revealed no noteworthy difference.

For measuring the transient signals, the gating chip was clocked with 15 MHz, which results in a duty cycle of 50% for a time duration of the active phase of the SPAD of ≈33.3 ns to be able to observe the whole discharging phase when an avalanche occurs. The minimum dead time corresponded to the time duration of the reset phase of the SPAD, which amounted to ≈33.3 ns. Figure 4 shows typical results of the cathode voltage's transient with and without avalanche events, where several active phases have been overlaid for illustration.

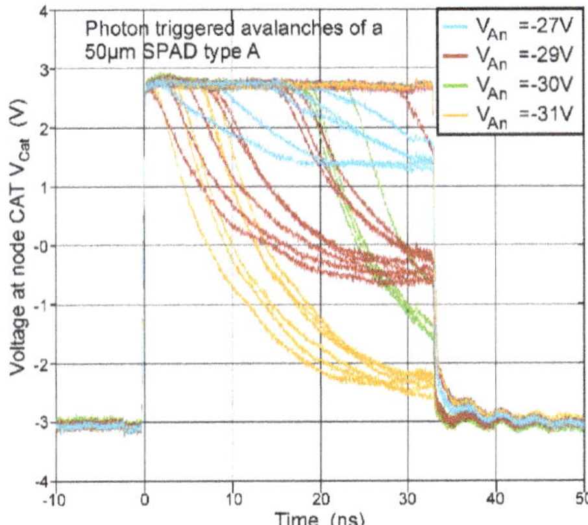

Figure 4. Transients of the cathode voltage (node CAT) of a SPAD type A with a diameter of 50 μm for different anode voltages V_{AN}. In the illustration, transients of several active phases are overlaid. It can be seen that early avalanches quench (themselves) during the active phase, when the breakdown level is reached. Some avalanches, which occurred later in the active phase are quenched by the following reset phase. The observed breakdown level in this figure depends on the anode voltage V_{AN}. It can be calculated with the breakdown voltage as $V_{AN} + V_{BD}$.

If an avalanche occurs, e.g., due to a photon hit, node CAT is discharged by the avalanche current of the SPAD. Avalanches, which happen earlier in the active phase have enough time to discharge node CAT until the breakdown level is reached. Later avalanches are quenched by the following reset phase. If no avalanche occurs, the voltage at node CAT remains and will be only switched below breakdown level during the reset phase.

With lots of samples of avalanche transients, which happened approximately in the first quarter of the active phase, an 80 to 20% fall time could be measured for the different SPADs. For this, the SPAD was illuminated with a halogen light so that the photon hits on the SPAD were somewhat equally distributed over the time duration of the active phase when overlaying several cycles with a storage oscilloscope as, e.g., depicted in Figure 5 for a SPAD type B with 48.2-μm diameter for an anode voltage of −66 V. When considering an ideal exponential decay of the excess bias, the time constant τ can be calculated by dividing the 80 to 20% fall time by ln(4).

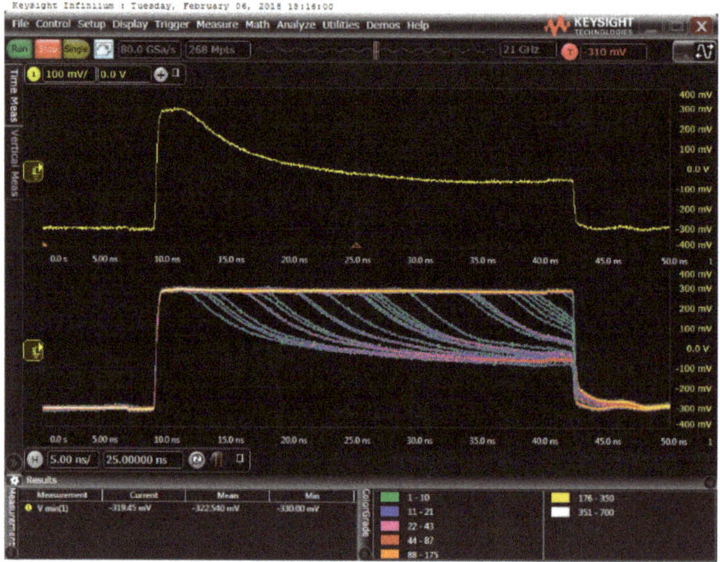

Figure 5. Oscilloscope picture of one avalanche (**top**) and several overlaid avalanches (**bottom**) of the cathode voltage of a SPAD type B with 48.2 μm diameter for an illumination power with a halogen source to achieve somewhat equally distributed photon hits.

Figure 6 shows the results for the different types of SPADs. There is a tendency that SPADs of type B in the HV CMOS technology are somewhat faster than SPADs of type A due to more avalanche current. The reason could be that for SPAD type B in HV CMOS technology deeper wells with a tendency to lower doping are used to meet requirements of high-voltage operation, which increases the breakdown voltage. This may result in a vertical thinner p epi layer (thinner absorption zone) below the SPAD type B, which in combination with the deep n-well increases somewhat the field in the depleted absorption zone. For SPADs with a diameter larger than 98.2 μm, a drop of the fall time was observed (meaning the avalanche build-up is faster) when the excess bias was increased, where the avalanche current got larger.

This was because the size of the SPADs' capacitance dominates over the pad capacitances. For the SPAD type A with 50 μm and SPAD type B with 48-μm diameter, the fall time remained approximately constant, because the pad and parasitic capacitances dominated. Concluding, the measurement shows a tendency to a lower fall time for larger SPADs.

For those avalanche transients, which could reach the breakdown level before the end of the active phase, the breakdown voltage V_{BD} was determined from the breakdown level V_{BL} from the relationship $V_{BD} = V_{BL}-V_{AN}$. In Figure 7, the results of the breakdown voltage for several SPADs versus the excess bias are depicted. All breakdown voltages were mostly independent from the excess bias (as expected). The small variation for SPAD type A with 400-μm diameter were originated from the gating chip, which came due to the larger capacitance of the SPAD to its operation limit. The measured breakdown voltages amounted to 28.5, 27.5, 26, and ≈27 V for SPAD type A with 50, 100, 200, and 400 μm, respectively. For SPAD type B breakdown voltages of 65.2 and 64.8 V, they were determined for diameters of 48.2 and 98.2 μm, respectively.

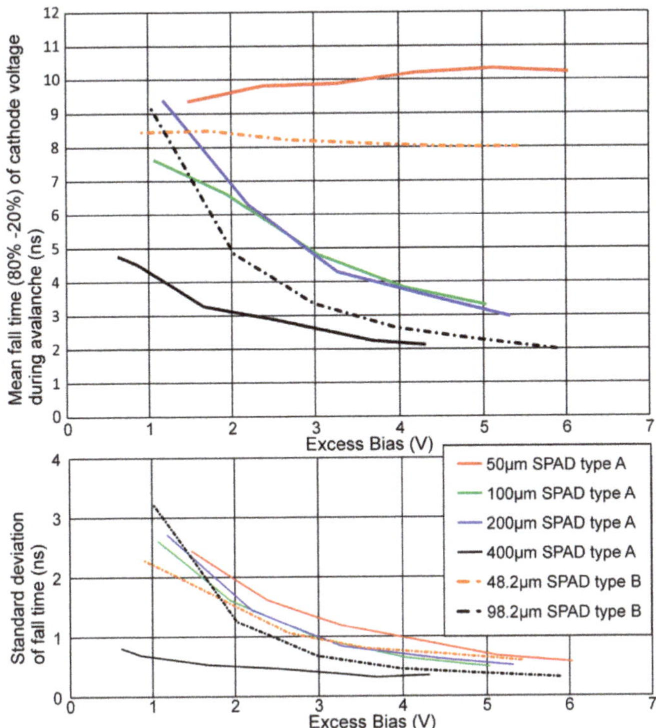

Figure 6. 80 to 20% fall time of the cathode voltage of different SPADs when an avalanche occurred.

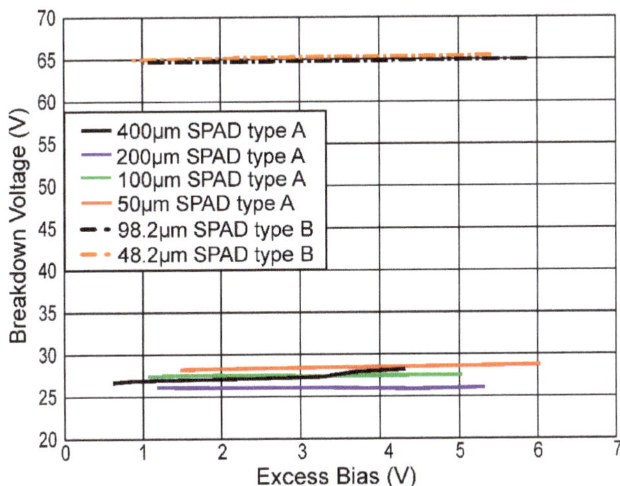

Figure 7. Breakdown voltage vs. excess bias.

The avalanche current transient can be calculated, if the overall node capacitance C_{Cat} is measured and the cathode voltage is evaluated by $C_{Cat} \times dV_{Cat}/dt$. The capacitances C_{Cat} for each SPAD, which was connected with a gating chip, were measured with an Agilent 4284A precision LCR meter. For SPAD type A, the anode voltage was chosen to be as close as possible below breakdown. Due to device restrictions, for SPAD type B, the anode voltage was set to −40 V for capacitance measurements. The results for the SPADs type A including pads and node CAT on the gating chip were 0.84, 1.12, 1.2, and 2.2 pF for the SPAD diameters of 50, 100, 200, and 400 μm, respectively. For SPAD type B, the capacitances 0.88 and 1.12 pF were measured for the diameters of 48.2 and 98.2 pF.

Figure 8 shows the resulting avalanche-current transients for SPAD type A with 200-μm diameter, and in Figure 9, the current transients for SPAD type B with 98.2 μm diameter are plotted. In Figures 8 and 9, the anode voltages V_{AN} of the SPADs are varied, which directly alters the excess bias. It can be seen, that at the beginning of the avalanche, the current raised towards a maximum current value, while the cathode of the SPAD was discharged and the excess bias dropped. This decrease operated against a further growing of the avalanche current, because the ionization coefficients for electrons and holes decline for lower field strengths. For every excess bias, which would be held constant, there exists a static avalanche current after breakdown. This avalanche current is smaller for a lower excess bias and vanishes at the breakdown level of the SPAD, where the excess bias amounts to zero. After the point of a peak avalanche-current, the current decreased, because the cathode's voltage dropped towards the breakdown level. The peak avalanche currents for all SPADs are plotted in Figure 10 in dependence on the excess bias.

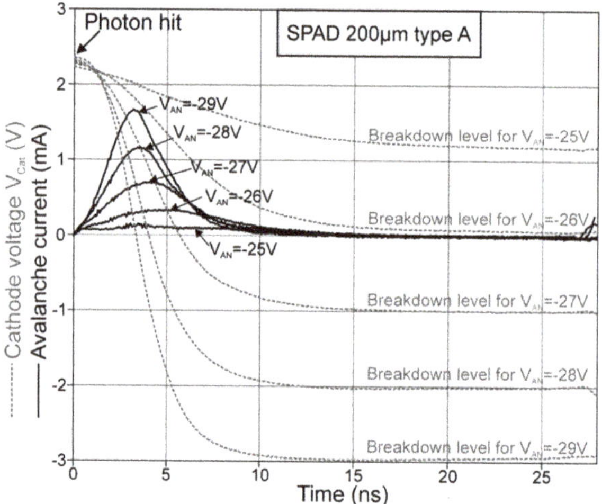

Figure 8. Avalanche-current transients for a SPAD type A with 200 μm diameter for different anode voltages V_{AN}, where the avalanche starts at 0 ns.

In Figure 8, there is a considerable shift of the peak current towards larger time points from 3 ns to almost 5 ns when changing the anode voltage from −29 to −26 V. For −25 V, noise influences the current transient. In Figure 9 the shift of the peak current occurs to a less extent. This can be explained, if the range of the excess bias from 0 to 6.6 V is seen in relation to the breakdown voltage, which was lower for the 200-μm SPAD type A with 26 V than for the 98.2-μm SPAD type B with 64.8 V. Therefore, the same excess bias caused a larger increase of the electric field in the multiplication region of SPAD type A than in SPAD type B.

For SPADs type A and SPADs type B, the peak avalanche current increased with raising excess bias and larger diameter. This can be observed in the same way in Figure 6, because a larger avalanche current caused a faster discharge and hence a lower fall time. For SPAD type A with a diameter of 50 μm and SPAD type B with a diameter of 48.2 μm, the pad capacitances were dominant and therefore the delay times and avalanche currents were nearly the same in the observed range of the excess bias.

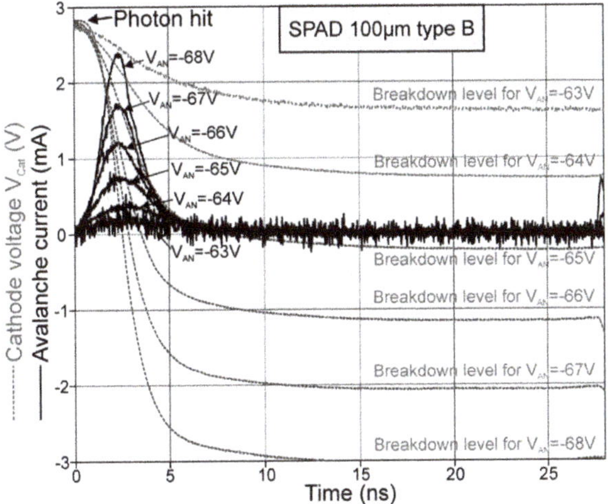

Figure 9. Avalanche-current transients for a SPAD type B with 98.2 µm diameter for different anode voltages V_{AN}, where the avalanche starts at 0 ns.

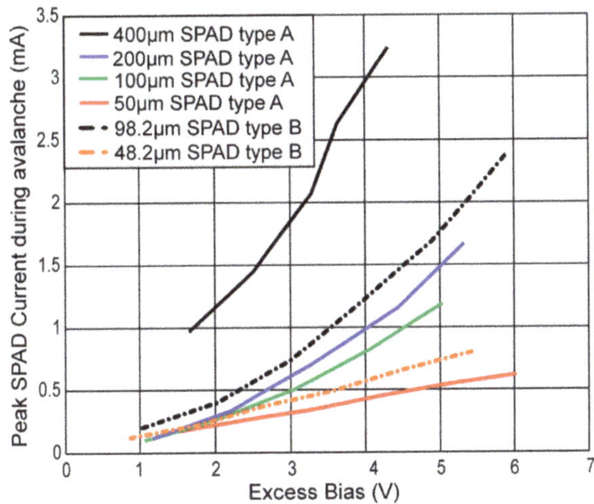

Figure 10. Peak avalanche-current for SPADs type A and B in dependence on the excess bias.

5. Conclusions

Knowing the behavior of SPADs helps to develop more detailed models. Especially for fully integrated data receivers, it is important to optimize the circuit to a distinct SPAD.

In this paper, the avalanche actions of two types of single-photon avalanche diodes (SPADs) in 0.35-µm CMOS technology were investigated. SPAD type A with diameters of 50, 100, 200, and 400 µm was designed in standard CMOS technology including a 12-µm thick p⁻ epi-layer. SPAD type B with diameters of 48.2 and 98.2 µm was designed in the high-voltage (HV) line of this technology. Each SPAD was wire-bonded to a clocked gating chip in standard 0.35-µm CMOS technology with a nominal supply voltage of 3.3 V. This chip controls in one clock period the charge up to maximum 6.6 V excess bias in the active phase as well as the quenching and read out in the reset phase. Measurements of the cathode voltage transients after photon hit at SPAD type A resulted in fall times (80 to 20%) of 10.2 ns for 50-µm SPAD diameter and an excess bias of 4.2 V and 3.45 ns for 200-µm SPAD diameter and an excess bias of 4.26 V. For type B fall times of 8 ns for 48.2-μm SPAD diameter and 5.4 V excess bias as well as 2 ns for 98.2-μm SPAD diameter and 5.9 V excess bias were determined. To calculate the avalanche current transients of the SPADs out of the transients of the cathode voltage in using the relationship $C_{Cat} \times dV_{Cat}/dt$, the whole capacitance at the cathode of the SPAD including gating chip connected were measured. This may be an alternative method to determine the current during discharging by the SPAD from a high-ohmic cathode node. It resulted in peak avalanche currents in the transients of, e.g., 1.19 mA for 100-μm SPAD type A and 1.64 mA for 98.2-μm SPAD type B for an excess bias of 5 and 4.9 V, respectively.

The breakdown voltages amounted to 28.5, 27.5, 26, and $\approx$27 V for SPAD type A with 50, 100, 200, and 400 μm, respectively, and 65.2 and 64.8 V for SPAD type B with diameters of 48.2 and 98.2 μm, respectively. Typically, the avalanche current rises for larger excess bias and larger diameter of the SPAD. Consequently, this has an effect on the fall time of the cathode voltage drop, when an avalanche occurs. The fall time depends on the avalanche current and typically gets smaller when the excess bias or the diameter of the SPAD is increased.

Author Contributions: Conceptualization, B.G., B.S., and H.Z.; methodology, B.G. and H.Z.; software, B.G.; validation, B.G. and H.Z.; formal analysis, B.G.; investigation, B.G.; resources, B.G., B.S., and H.Z.; data curation, B.G.; writing—original draft preparation, B.G.; writing—review and editing, B.G and H.Z.; visualization, B.G.; supervision, H.Z.; project administration, H.Z.; funding acquisition, H.Z. All authors have read and agreed to the published version of the manuscript.

Funding: This research was funded by AUSTRIAN SCIENCE FOUNDATION (FWF), project P28335-N30.

Acknowledgments: The authors thank the Austrian Science Foundation (FWF) for financial funding in the project P28335-N30.

Conflicts of Interest: The authors declare no conflict of interest. The funders had no role in the design of the study; in the collection, analyses, or interpretation of data; in the writing of the manuscript, or in the decision to publish the results.

References

1. Charbon, E.; Fishburn, M.W. Monolithic Single-Photon Avalanche Diodes: SPADs. In *Single Photon Imaging*; Seitz, P., Theuwissen, A.J.P., Eds.; Springer Series in Optical Sciences; Springer: Heidelberg, Germany; Dordrecht, The Netherlands; London, UK; New York, NY, USA, 2011; pp. 123–157.

2. Itzler, M.A.; Jiang, X.; Entwistle, M.; Slomkowski, K.; Tosi, A.; Acerbi, F.; Zappa, F.; Cova, S. Advances in InGaAsP-based avalanche diode single photon detectors. *J. Mod. Opt.* **2011**, *58*, 174–200. [CrossRef]

3. Itzler, M.A.; Jiang, X.; Entwistle, M.; Slomkowski, K.; Tosi, A.; Acerbi, F.; Zappa, F.; Cova, S. Afterpulsing Effects in Free-Running InGaAsP Single-Photon Avalanche Diodes. *IEEE J. Quantum Electron.* **2008**, *44*, 3–11.

4. Webster, E.A.G.; Grant, L.A.; Henderson, R.K. A High-Performance Single-Photon Avalanche Diode in 130-nm CMOS Imaging Technology. *IEEE Electron. Device Lett.* **2011**, *33*, 1580–1591. [CrossRef]

5. Hofbauer, M.; Steindl, B.; Zimmermann, H. Temperature Depencence of Dark Count Rate and After Pulsing of a Single-Photon Avalanche Diode with an Integrated Active Quenching Circuit in 0.35 μm CMOS Technology. *Hindawi J. Sens.* **2018**, *2018*, 9585931.

6. Sanzaro, M.; Gattari, P.; Villa, F.; Tosi, A.; Croce, G.; Zappa, F. Single-Photon Avalanche Diodes in a 0.16 μm BCD Technology with Sharp Timing Response and Red-Enhanced Sensitivity. *IEEE J. Sel. Top. Quantum Electron.* **2018**, *24*, 3801209. [CrossRef]

7. Cavaliere, F.; Prati, E.; Poti, L.; Muhammad, I.; Catuogno, T. Secure Quantum Communication Technologies and Systems: From Labs to Market. *Quantum Rep.* **2020**, *2*, 80–106. [CrossRef]

8. Ding, Y.; Bacco, D.; Llevellyn, D.; Faruque, I.; Paesani, S.; Galili, M.; Liang, A.; Rottwitt, K.; Thompson, M.; Wang, J.; et al. Silicon Photonics for Quantum Communications. In Proceedings of the 21st International Conference on Transparent Optical Networks (ICTON), Angers, France, 9–13 July 2019; p. We.E5.4.

9. Acerbi, F.; Bisadi, Z.; Fontana, G.; Zorzi, N.; Piemonte, C.; Pavesi, L. A Robust Quantum Random Number Generator Based on an Integrated Emitter-Photodetector Structure. *IEEE J. Sel. Top. Quantum Electron.* **2018**, *24*, 161–167. [CrossRef]

10. Tontini, A.; Gasparini, L.; Massari, N.; Passerone, R. SPAD-Based Quantum Random Number Generator With an Nth-Order Rank Algorithm on FPGA. *IEEE Trans. Circuits Syst. II* **2019**, *66*, 2067–2071. [CrossRef]

11. Khanmohammadi, A.; Enne, R.; Hofbauer, M.; Zimmermann, H. A Monolithic Silicon Quantum Random Number Generator based on Measurement of Photon Detection Time. *IEEE Photonics J.* **2015**, *7*, 1–13. [CrossRef]

12. Portaluppi, D.; Conca, E.; Villa, F. 32 × 32 CMOS SPAD Imager for Gated Imaging, Photon Timing, and Photon Coincidence. *IEEE J. Sel. Top. Quantum Electron.* **2018**, *24*, 3800706. [CrossRef]

13. Jahromi, S.; Jansson, J.-P.; Keränen, P.; Kostamovaara, J. A 32 × 128 SPAD-257 TDC Receiver IC for Pulsed TOF Solid-State 3-D Imaging. *IEEE J. Solid-State Circuits* **2020**, *55*, 1960–1970. [CrossRef]

14. Della Rocca, F.M.; Mai, H.; Hutchings, S.W.; Al Abbas, T.; Buckbee, K.; Tsiamis, A.; Lomax, P.; Gyongy, I.; Dutton, N.A.; Henderson, R.K. A 128 × 128 SPAD Motion-Triggered Time-of-Flight Image Sensor with In-Pixel Histogram and Column-Parallel Vision Processor. *IEEE J. Solid-State Circuits* **2020**, *55*, 1762–1775. [CrossRef]

15. Henderson, R.K.; Johnston, N.; Hutchings, S.W.; Gyongy, I.; Al Abbas, T.; Dutton, N.; Tyler, M.; Chan, S.; Leach, J. A 256 × 256 40 nm/90 nm CMOS 3D-Stacked 120dB-Dynamic-Range Reconfigurable Time-Resolved SPAD Imager. In Proceedings of the IEEE International Solid-State Circuits Conference (ISSCC), San Francisco, CA, USA, 17–21 February 2019; pp. 106–107.

16. Choi, J.; Taal, A.J.; Pollmann, E.H.; Lee, C.; Kim, K.; Moreaux, L.C.; Roukes, M.L.; Shepard, K.L. A 512-Pixel, 51-kHz-Frame-Rate, Dual-Shank, Lens-Less, Filter-Less Single-Photon Avalanche Diode CMOS Neural Imaging Probe. *IEEE J. Solid-State Circuits* **2019**, *54*, 2957–2968. [CrossRef] [PubMed]

17. Perenzoni, M.; Massari, N.; Gasparini, L.; Garcia, M.M.; Perenzoni, D.; Stoppa, D. A Fast 50 × 40-Pixels Single-Point DTOF SPAD Sensor With Photon Counting and Programmable ROI TDCs, With σ <4 mm at 3 m up to 18 klux of Background Light. *IEEE Solid-State Circuits Lett.* **2020**, *3*, 86–89.

18. Zhang, L.; Chitnis, D.; Chun, H.; Rajbhandari, S.; Faulkner, G.; O'Brien, D.; Collins, S. A Comparison of APD- and SPAD-Based Receivers for Visible Light Communications. *IEEE J. Lightwave Technol.* **2018**, *36*, 2435–2442. [CrossRef]

19. Fisher, E.; Underwood, I.; Henderson, R. A Reconfigurable Single-Photon-Counting Integrating Receiver for Optical Communications. *IEEE J. Solid-State Circuits* **2013**, *48*, 1638–1650. [CrossRef]

20. Zimmermann, H.; Steindl, B.; Hofbauer, M.; Enne, R. Integrated fiber optical receiver reducing the gap to the quantum limit. *Sci. Rep.* **2017**, *7*, 2652. [CrossRef]

21. Steindl, B.; Hofbauer, M.; Schneider-Hornstein, K.; Brandl, P.; Zimmermann, H. Single-Photon Avalanche Photodiode Based Fiber Optic Receiver for Up to 200 Mb/s. *IEEE J. Sel. Top. Quantum Electron.* **2018**, *24*, 3801308. [CrossRef]

22. *Forward Error Correction for High Bit Rate DWDM Submarine Systems*; ITU-Recommendation G. 975.1; International Telecommunication Union: Geneva, Switzerland, 2004.

23. Kosman, J.; Almer, O.; Al Abbas, T.; Dutton, N.; Walker, R.; Videv, S.; Moore, K.; Haas, H.; Henderson, R. A 500 Mb/s −46.1 dBm CMOS SPAD Receiver for Laser Diode Visible-Light Communications. In Proceedings of the IEEE International Solid-State Circuits Conference (ISSCC), San Francisco, CA, USA, 17–21 February 2019; pp. 468–469.

24. Goll, B.; Hofbauer, M.; Steindl, B.; Zimmermann, H. A Fully Integrated SPAD-Based CMOS Data-Receiver with a Sensitivity of −64 dBm at 20 Mb/s. *IEEE Solid-State Circuits Lett.* **2018**, *1*, 2–5. [CrossRef]

25. Goll, B.; Hofbauer, M.; Steindl, B.; Zimmermann, H. Transient Response of a 0.35 μm CMOS SPAD with Thick Absorption Zone. In Proceedings of the 25th IEEE International Conference on Electronics, Circuits and Systems (ICECS), Bordeaux, France, 9–12 December 2018; pp. 9–12.

26. Steindl, B.; Enne, R.; Zimmermann, H. Tick detection zone single-photon avalanche diode fabricated in 0.35 μm complementary metal-oxid semiconductors. *SPIE Opt. Eng. Lett.* **2015**, *54*, 050503.

27. Enne, R.; Steindl, B.; Hofbauer, M.; Zimmermann, H. Fast Cascoded Quenching Circuit for Decreasing Afterpulsing Effects in 0.35-μm CMOS. *IEEE Solid-State Circuits Lett.* **2018**, *1*, 62–65. [CrossRef]

28. Hofbauer, M.; Steindl, B.; Schneider-Hornstein, K.; Zimmermann, H. Performance of high-voltage CMOS single-photon avalanche diodes with and without well-modulation technique. *SPIE Opt. Eng.* **2020**, *59*, 040502. [CrossRef]

29. Hofbauer, M.; Steindl, B.; Schneider-Hornstein, K.; Zimmermann, H. Thick Single-Photon Avalanche Diode Optimized for Near Infrared with Integrated Active Quenching Circuit. In Proceedings of the Single Photon Workshop, Milano, Italy, 21–25 October 2019; p. 150.

30. Bronzi, D.; Villa, F.; Tisa, S.; Tosi, A.; Zappa, F. SPAD Figures of Merit for Photon-Counting, Photon-Timing and Image Applications: A Review. *IEEE Sens. J.* **2016**, *16*, 3–12. [CrossRef]

Article

Metal-Semiconductor-Metal GeSn Photodetectors on Silicon for Short-Wave Infrared Applications

Soumava Ghosh [1,2,3], **Kuan-Chih Lin** [2], **Cheng-Hsun Tsai** [2], **Harshvardhan Kumar** [1,2], **Qimiao Chen** [4], **Lin Zhang** [4], **Bongkwon Son** [4], **Chuan Seng Tan** [4], **Munho Kim** [4], **Bratati Mukhopadhyay** [3] and **Guo-En Chang** [1,2,*]

1   Department of Mechanical Engineering, and Advanced Institute of Manufacturing with High-Tech Innovations (AIM-HI), National Chung Cheng University, Chiyai County 62102, Taiwan; ghoshsoumava2@gmail.com (S.G.); harshvardhan@nitdelhi.ac.in (H.K.)
2   Graduate Institute of Opto-Mechatronics, National Chung Cheng University, Chiyai County 62102, Taiwan; xavierlin0105@gmail.com (K.-C.L.); a0918663250@gmail.com (C.-H.T.)
3   Institute of Radio Physics and Electronics, University of Calcutta, Kolkata 700009, India; bratmuk@yahoo.co.in
4   School of Electrical and Electronic Engineering, Nanyang Technological University, Singapore 639798, Singapore; chenqm@ntu.edu.sg (Q.C.); zhang.lin@ntu.edu.sg (L.Z.); BONGKWON001@e.ntu.edu.sg (B.S.); TanCS@ntu.edu.sg (C.S.T.); munho.kim@ntu.edu.sg (M.K.)
*   Correspondence: imegec@ccu.edu.tw; Tel.: +886-5-2720411 (ext. 33324)

Received: 25 July 2020; Accepted: 19 August 2020; Published: 21 August 2020

Abstract: Metal-semiconductor-metal photodetectors (MSM PDs) are effective for monolithic integration with other optical components of the photonic circuits because of the planar fabrication technique. In this article, we present the design, growth, and characterization of GeSn MSM PDs that are suitable for photonic integrated circuits. The introduction of 4% Sn in the GeSn active region also reduces the direct bandgap and shows a redshift in the optical responsivity spectra, which can extend up to 1800 nm wavelength, which means it can cover the entire telecommunication bands. The spectral responsivity increases with an increase in bias voltage caused by the high electric field, which enhances the carrier generation rate and the carrier collection efficiency. Therefore, the GeSn MSM PDs can be a suitable device for a wide range of short-wave infrared (SWIR) applications.

Keywords: photodetectors; GeSn alloys; silicon photonics; photonic integrated circuits

1. Introduction

The monolithic integration of photonic components and devices on a silicon (Si) platform to create electronic–photonic integrated circuits (EPICs) has attracted immense research interest over recent decades [1–5]. The key driving force is the bottleneck in the data transport rate due to the limited data rates of~10 GB/s of the current copper interconnects. Nowadays, optical and photonic interconnects operating in the fiber-optic low-loss windows (1264–1675 nm) of silica fibers have been proposed to replace metal interconnects to increase the data transport rate and reduce thermal problems. Among the various active photonic components, photodetectors (PDs) are a crucial building block. Although III-V semiconductor compounds are mostly used for high-speed photodetection, these compounds are not compatible with the Si complementary metal-oxide-semiconductor (CMOS) IC technology [6].

Alternatively, the compatibility with Si-based CMOS processing technology, monolithic integration on the same Si chip [7], and low fabrication cost make group-IV materials very attractive to develop CMOS-compatible photonic devices. However, the realization of those devices for the most important fiber-optical communication window (1310 and 1550 nm) by Si is not possible due to its bandgap of 1.12 eV [8], resulting in a cutoff wavelength of ~1100 nm. This problem can be partially circumvented by using Ge, as its direct bandgap (0.8 eV) supports 1310 nm at room temperature. However, beyond 1500 nm wavelength, the responsivity of Ge falls drastically [9]. Therefore, the entire telecommunication bands cannot be covered by Ge based photodetectors (PDs).

Over recent decades, the growth of high-quality $Ge_{1-x}Sn_x$ thin film by chemical vapor deposition (CVD) and molecular beam epitaxy (MBE) [10–12] on Si substrate via a suitable buffer has opened new avenues for group-IV photonics. The incorporation of Sn in Ge not only modifies the electronic band structure by shrinking the direct bandgap and hence red-shifting the absorption edge [13], but also beyond 8% Sn concentration the GeSn alloy acts as a direct bandgap semiconductor [14]. This noteworthy feature of GeSn has encouraged researchers to develop different types of optoelectronic devices such as Light Emitting Diodes (LEDs) [15–17], LASERs [18–20], Transistor LASERs [21,22], *p-i-n* PDs [23–25], quantum well infrared photodetectors (QWIPs) [26–28], metal-semiconductor-metal photodetectors (MSM PDs) [29–31], waveguide PDs [32], and heterojunction bipolar phototransistors (HPTs) [33–40].

MSM PDs are an alternative choice of *p-i-n* PD, consisting of back-to-back Schottky diodes. As MSM PDs do not require any doping, the effect of parasitic capacitance cannot degrade the performance [41]. Only transit-time limited delay presents; therefore, the operation speed is higher than normal *p-i-n* PDs. The simple planar fabrication of MSM PDs is suitable for monolithic integration with other components of photonic circuits [42,43]. Yasar et al. [29] and Mahmodi et al. [30] reported amorphous (8% Sn content) and crystalline GeSn MSM PD on Si substrates, respectively, but they only focused on the measurement of dark and photocurrent. Recently, the responsivity of $Ge_{1-x}Sn_x$ thin-film based MSM PD on Si has been reported [31]; however, in their work, they only demonstrated up to 1000 nm wavelength which cannot cover the modern telecommunication window (1550 nm).

In this work, we demonstrate the GeSn MSM PD on the Si platform for efficient photodetection in the entire telecommunication bands. We show the material growth and electrical and optical characterization results of the fabricated GeSn MSM PD with 4% Sn content. Furthermore, we also measured the spectral responsivity for different bias voltages and analyzed the strain electronic band structure and absorption to study the enhanced photodetection.

2. Materials and Methods

2.1. Device Design

Figure 1a exhibits the 3D schematic diagram of the designed surface-illuminated GeSn MSM PD. The layer structure consisted of a GeSn active layer as the absorption layer grown on the Si substrate via Ge virtual substrate (VS). The presence of GeSn active region helped to enhance the absorption capacity with respect to the pure Ge, due to the smaller direct bandgap and the larger absorption coefficient. The GeSn layer was passivated by the SiO_2 layer. Two metal pads were deposited on the top surface of the GeSn active layer. A schematic band diagram is shown in Figure 1b.

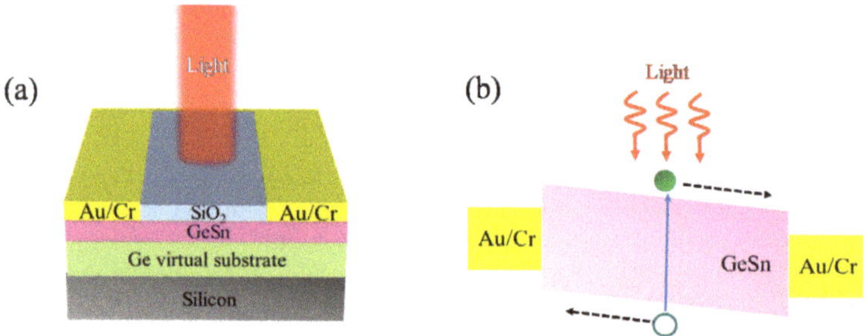

Figure 1. (a) 3D Schematic diagram of our designed surface-illuminated GeSn metal-semiconductor-metal photodetectors (MSM PDs) on Ge buffered Si substrates. (b) Schematic band diagram of the GeSn MSM PD.

As light is normally incident on the device and absorbed by the GeSn active layer, electrons and holes could be generated in the GeSn active layer. In the presence of the bias voltage, the potential difference between positive and negative electrical contacts created a band bending. Under the illuminated conditions, the electron-hole pairs were generated in the active region and then swept out to the electrodes in the presence of the induced electric field. The holes and electrons accumulated in the negative and positive metal contacts, respectively, resulting in the flow of the photocurrent. The materials and dimensions of the different regions are listed in Table 1.

Table 1. Materials and dimensions of the GeSn metal-semiconductor-metal photodetectors (MSM PDs).

Layer	Material	Thickness (nm)
Substrate	Si	150
Virtual Substrate	Ge	900
Active Layer	$Ge_{0.96}Sn_{0.04}$	180
Top Passivation Layer	SiO_2	180

2.2. Material Growth and Characterization

The GeSn sample was grown on a 150 mm Si (001) substrate in industry compatible reduced pressure-chemical vapor deposition (RP-CVD) chamber from ASM by using Ge_2H_6 and $SnCl_4$ as the precursors. Prior to the deposition of the GeSn layer, a high-quality Ge VS with a thickness of ~900 nm was deposited on the Si substrate by using Ge_2H_6 as the precursor at 400 °C followed by annealing at 850 °C for 30 min. After that, the temperature in the chamber was decreased to 325 °C, and a 180 nm thick of GeSn film was deposited on Ge VS as the active layer. The Sn concentration and strain of the GeSn layer were characterized at room temperature by X-ray diffraction reciprocal space mapping (XRDRSM) using a PANalytical X'Pert diffractometer. The microstructure of the GeSn sample was investigated by cross-sectional transmission electron microscopy (XTEM) in dark field mode (FEI Tecnai G2 F20, FEI, Waltham, MA, USA). A Pt layer was deposited on the surface of the GeSn sample to increase the conductivity for XTEM experiments.

2.3. MSM PD Fabrication

The surface-illuminated GeSn MSM PDs were fabricated using a standard CMOS-compatible process. A square mesa with a width of 1 mm was created by standard optical lithography followed by reactive ion etching (RIE) techniques. Next, a 180 nm thick SiO_2 passivation layer was deposited using plasma-enhanced chemical vapor deposition (PECVD), which acts as an electric isolator between positive and negative electrodes and an anti-reflection (AR) layer to enhance the optical responsivity. After that, contact windows were incorporated using optical lithography and wet etching methods with buffered oxide etch (BOE) solution to expose the semiconductor surfaces for making the electrical contacts. Finally, 200/20 nm thick Au/Cr metal pads were deposited using an e-beam evaporator and patterned rectangular-shaped using a lift-off process.

2.4. Electrical and Optical Measurements

A Keithley 2400 SourceMeter was used to characterize the electrical properties of the fabricated GeSn MSM PDs at room temperature under dark and illuminated conditions. To characterize the spectral property of the GeSn MSM PDs, Fourier-transform infrared spectroscopy (FTIR) was used. The emitted light from the FTIR was incident normally on the devices. A Keithley 2400 SourceMeter and a 50 Ω load resistance in series were used to bias the GeSn MSM PDs. The voltage drop across the load resistance was fed back to the FTIR for the determination of photocurrents. The responsivity of the GeSn MSM PDs was then obtained by calibrating the optical responses using a commercial extended InGaAs PD (Thorlabs DET10D2, Thorlabs, Inc., Newton, NJ, USA) to determine the optical responsivity.

3. Results and Discussion

3.1. Material Characterization

Figure 2a shows a cross-sectional XTEM of the grown GeSn sample. Most defects were confined in the region near the Ge/Si interface, and no obvious defects were found near the surface of the Ge VS. The single-crystalline GeSn had a thickness of ~180 nm which was below the critical thickness for plastic relaxation, resulting in a high-quality GeSn layer. As a result, no obvious threading defects were observed in the TEM image of the GeSn layer, as shown in Figure 2a, suggesting the GeSn layer was pseudomorphic to the underlying Ge VS. The threading dislocation density (TDD) of the GeSn layer was estimated to be ~4.7×10^7 cm^{-2} by etch pit density (EPD) methods. Figure 2b shows a (224) XRDRSM of the grown GeSn sample, from which three peaks were observed, associated with the Si substrate, Ge VS, and GeSn layer. From the peak positions, the concentration and strain could be extracted. In addition, the Ge and GeSn peaks were aligned at the same Q_x, suggesting that GeSn was pseudomorphically grown on Ge VS. The diagonal line through the Si peak indicates that Ge VS was almost fully relaxed (~0.1% tensile strain due to the annealing process). The Sn concentration and compressive strain of the GeSn layer were determined to be 4% and 0.57%, respectively.

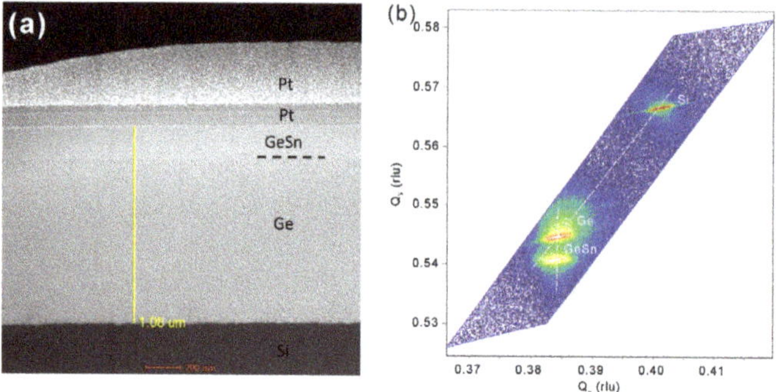

Figure 2. (**a**) Cross-sectional transmission electron microscopy (XTEM) image of the grown GeSn/Ge/Si sample. (**b**) (224) X-ray diffraction reciprocal space mapping (XRDRSM) of the GeSn/Ge/Si sample, revealing a pseudomorphic heterostructure.

To confirm the bandgap of the grown GeSn sample, photoluminescence (PL) experiments were performed at room temperature using a 532 nm laser as the light source. The emitted PL signals from the GeSn sample were recorded uisng Fourier Transform infrared spectroscopy (FTIR) with ab LN2-cooled InSb photodetector. Figure 3 shows the measured room-temperature PL spectrum from the grown GeSn sample. Although there was signal distortion in the range of 1700–1900 nm due to the atmospheric absorption, a single emission peak was observed near 1800 nm. In addition, the emission peak was asymmetrical, confirming direct-bandgap light emission. The measured PL spectrum was modeled using the modified Gaussian function to obtain the emission peak position; the results are depicted in Figure 3. From the results, the emission peak was determined to be 1810 nm. The emission peak ($E_{\max}$) was related to the direct bandgap energy E_g^Γ via

$$E_{\max} = E_g^\Gamma + \frac{kT}{2} \tag{1}$$

where k is the Boltzmann constant and T is temperature. Using the $E_{\max}$ obtained from the room-temperature PL spectrum, we obtained $E_{\max} = 0.672$ eV. This value was obviously smaller than that of pure Ge (0.8 eV), showing the reduced direct bandgap of the GeSn materials due to Sn-alloying.

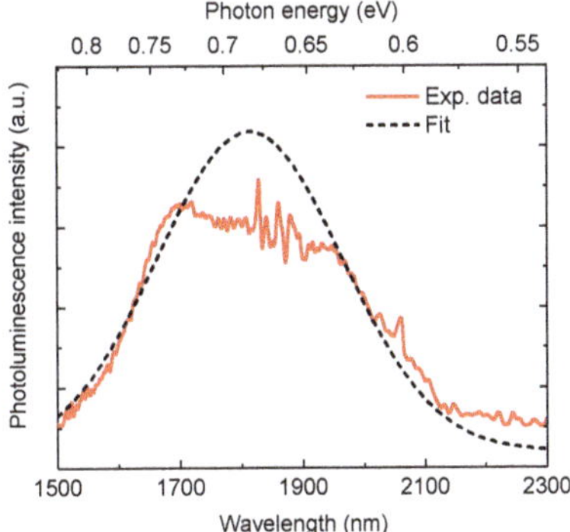

Figure 3. Room-temperature photoluminescence spectrum of the grown GeSn/Ge/Si sample. The signal distribution in the range of 1700–1900 nm is attributed to the atmospheric absorption.

3.2. Dark Current, Photogenerated Current, and Gains

Figure 4a shows the current-voltage (I-V) characteristics under dark and illumination conditions. The symmetric nature of the dark current at forward and reverse bias was observed due to the design of the symmetrical electrodes. In addition, the I-V curves were also seen to be nonlinear, this indicated the presence of the Schottky contact between the metal and GeSn layer. The increase in bias voltage increased the band to band tunneling mechanisms; therefore, the dark current increased. In addition, our fabricated GeSn-based MSM PD exhibited a lower measured value of dark current compared to the fabricated Ge-based MSM PD [44], suggesting good material quality. Under illumination with a 1510 nm laser source and an optical power of 7.7 mW, the enhancement of the current confirmed the photodetection ability of the fabricated device. Figure 4b exhibits the current gain which could be obtained from the ratio of current under illumination ($I_{Illumination}$) and dark current (I_{dark}) as a function of bias voltage. The current gain showed a constant nature with applied bias due to the constant enhancement of the current under dark and illumination.

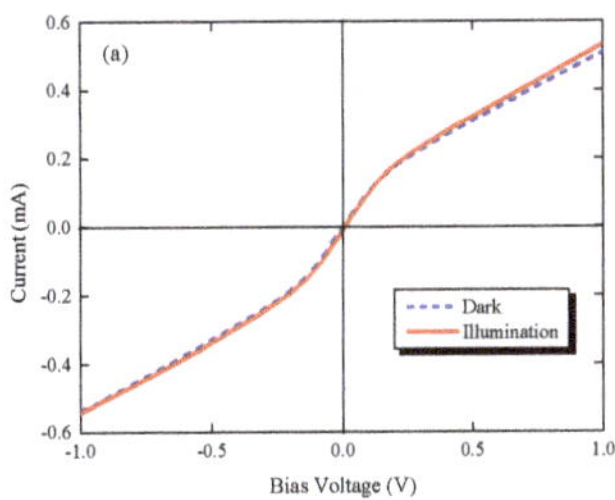

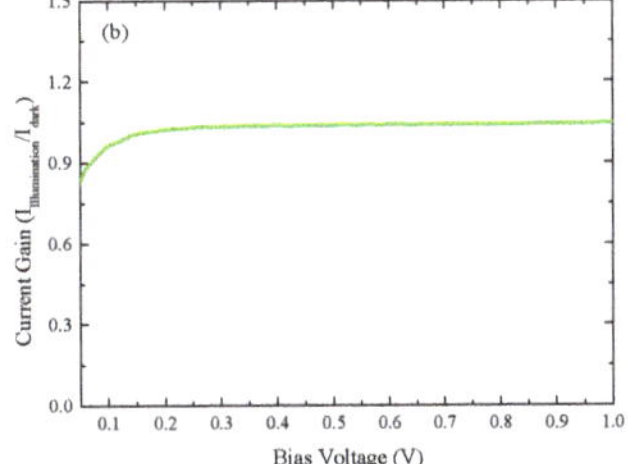

Figure 4. (a) Measured dark current and photocurrent variation with bias voltage at $T = 300$ K. (b) Measured current gain ($I_{Illumination}/I_{dark}$) variation with bias voltage at $T = 300$ K.

3.3. Spectral Responsivity

To obtain the maximum spectral responsivity, a SiO_2 AR layer was employed to reduce the reflection of incident light for MSM PDs. Figure 5a shows the measured reflectivity for the GeSn MSM PDs. For the GeSn MSM PDs, reflectivity showed low values of 20%–30% in the wavelength of 1400 to 2000 nm. The ripple-like features of reflectivity spectra were observed due to the interference between the layers. Figure 5b shows the measured responsivity spectra of the GeSn MSM PDs with different bias voltages at room temperature and the reflectivity spectrum. It can be seen that the responsivity of the device decreased with an increase in the wavelength. From the responsivity at a bias voltage of 1V, the cutoff-wavelength was estimated to be 1800 nm. This extended photodetection cutoff wavelength compared to 1550 nm of Ge PDs was caused by the incorporation of Sn in the active layer, indicating that the lowest direct bandgap of the GeSn active layer was $E_g^\Gamma = 688$ meV, which is in reasonable agreement with the PL results. The measured results suggested that the photodetection range of our devices entirely covered the telecommunication O-, E-, S-, C-, L-, and U- bands; thus, it is useful for telecommunication applications. Beyond 1800 nm, the optical responsivity was low because only inefficient indirect-gap interband absorption contributed to the optical absorption. Furthermore, the spectral responsivity of the device increased with increasing bias voltage. The enhancement ratios for GeSn MSM PDs for 1, 3, 5, and 7 V were about 84.5%, 219.6%, and 369.6%, respectively, compared to the referential PD for 1V at 1550 nm. The increase in responsivity with bias voltage could be explained based on the high electric field, which enhanced the carrier generation rate and the carrier collection efficiency. However, a high bias voltage caused more power dissipation in the device; therefore, the low bias voltage was preferred for the operation of the device. The applied voltage could be significantly reduced by shrinking the size of the GeSn MSM, while a high responsivity could be maintained. The designed device showed the peak responsivity of 40 mA/W at 1550 nm, which is much higher than that of conventional SiGe-based MSM PDs [45–47], showing the unique advantages of GeSn MSM PDs for telecommunication applications. The high spectral responsivity of GeSn MSM PD was due to the direct bandgap nature, high absorption coefficient, and high carrier mobility of GeSn alloy in the active region. In addition, it was also observed that the optical responsivity beyond 1800 nm also significantly increased with increasing applied bias voltage. This observation could be attributed to the Joule heating effect that enhanced lattice vibrations (phonons), resulting in enhanced indirect-gap absorption. Further enhancement in optical responsivity is possible by increasing the Sn content to further increase the absorption coefficient and/or optimizing the device structure to increase the carrier collection efficiency to enable more sensitive short-wave infrared (SWIR) photodetection.

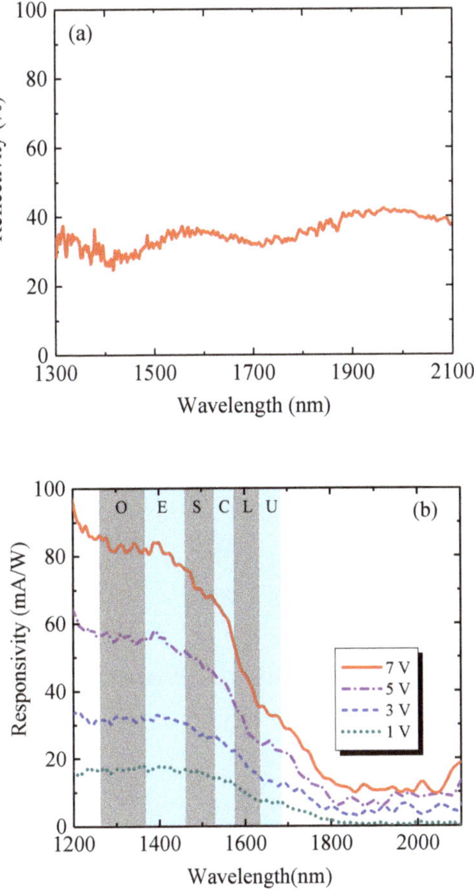

Figure 5. (**a**) Measured reflectivity spectra for the GeSn MSM PDs. (**b**) Measured responsivity spectra of the GeSn MSM PDs with different bias voltages at T = 300 K.

3.4. Numerical Analysis

The incorporation of Sn into Ge led to the extension of the absorption edge to the higher wavelength. This was because of the reduction in the direct bandgap of the GeSn alloy with Sn alloying. With an increase in Sn concentration, the Γ-conduction band shifted downward mainly due to the negative bandgap of $\alpha - Sn$ and the bowing effect of the direct bandgap. However, the heavy-hole (HH) band light-hole (LH) band in the valence band shifted upwards with an increase in Sn concentration. Therefore, direct-gap interband transitions contributed to the increased absorption spectra with Sn alloying owing to the increased density-of-states (DOS). To obtain a clear visualization of the effect of Sn concentration on the absorption coefficient, we theoretically calculated the strained electronic band structures using the deformation potential theory [42,48]. Then, the direct band absorption coefficient was calculated by using the Fermi's golden rule with consideration of a Lorentzian lineshape function [25,49],

$$\alpha(h\omega) = \frac{\pi he^2}{n_r c \varepsilon_0 m_0^2 h\omega} \sum_m \int \frac{2d\mathbf{k}}{(2\pi)^3} |\hat{e} \cdot p_{CV}|^2 \times \frac{\gamma/(2\pi)}{[E_{C\Gamma}(\mathbf{k}) - E_m(\mathbf{k}) - h\omega]^2 + (\gamma/2)^2} \tag{2}$$

where n_r is the refractive index; c is the velocity of light in free space; e is the electronic charge; $\hbar$ is the reduced Planck's constant; m_0 is the rest mass of an electron; ε_0 is the free space permittivity; ω is the angular frequency of incident light; $|\hat{e} \cdot p_{CV}|^2$ is the momentum matrix; γ is the full-width-at-half-maximum (FWHM) of the Lorentzian lineshape, whose value 15 meV was used in this study; $E_{C\Gamma}(\mathbf{k})$ and $E_m(\mathbf{k})$ are the electron and hole energy in the Γ-valley conduction band (CB) and valance band (VB), respectively, which were calculated using a multi-band k·p method by considering the strain effect [33,42]. The summation over m represents all interband transitions from the VB (HH and LH bands) to the direct CB. For the indirect-band absorption, because the probability of the indirect-gap transition was much smaller than that of the direct-gap transition, we neglected the indirect-band absorption effect in this study. The parameters for GeSn alloys could be evaluated from the linear interpolation of the Ge and Sn, as shown in Table 2. The direct bandgap bowing parameter of GeSn alloy was b_Γ = 2.42 eV [33].

Table 2. Parameters for Ge and Sn at T = 300 K.

Parameters	Ge	Sn
Lattice Constant a (Å)	5.6573 [19]	6.4892 [19]
Electron Effective Masses		
Electron m_Γ/m_0	0.038 [19]	0.058 [19]
Heavy Hole m_{HH}/m_0	0.28 [34]	
Light Hole m_{LH}/m_0	0.044 [34]	
Band gap $E_{g\Gamma}$ (eV)	0.7985 [19]	−0.413 [19]
Spin-orbit splitting Energy Δ_0 (eV)	0.29 [19]	0.80 [19]
Average Valance Band Energy E_{VaV} (eV)	0 [19]	0.69 [19]
Luttinger's parameters		
γ_1	13.38 [19]	−14.97 [19]
γ_2	4.24 [19]	−10.61 [19]
γ_3	5.69 [19]	−8.52 [19]
Deformation Potential		
a_c (eV)	−8.24 [19]	−6.00 [19]
a_v (eV)	1.24 [19]	1.58 [19]
b_v (eV)	−2.90 [19]	−2.70 [19]
Elastic Constants		
C_{11} (GPa)	128.53 [19]	69.00 [19]
C_{12} (GPa)	48.26 [19]	29.30 [19]
C_{44} (GPa)	68.30 [19]	36.20 [19]
Electron mobility μ_n (cm^2/V-sec)	3900 [28]	2940 [28]
Hole mobility μ_p (cm^2/V-sec)	1900 [28]	2990 [28]
Optical Energy E_p (eV)	26.3 [19]	24.0 [19]
Refractive Index n_r	4.051 [28]	5.791 [28]
Dielectric Constant ε_r	16.2 [28]	24.0 [28]

Figure 6a shows a schematic band diagram of pseudomorphic Ge$_{0.96}$Sn$_{0.04}$ on Ge with respect to wavenumber (k)at T = 300 K. The pseudomorphic growth of GeSn with 4% Sn on Ge VS exerted a compressive strain (~0.57%) which split the degeneracy of the HH and the LH bands. It is clearly shown from Figure 6a that the HH band shifted above the LH band, and their separation energy became larger with increasing Sn concentration due to the larger compressive strain. Therefore, two possible direct interband transitions contributed to the optical absorption: the direct transition from the HH band to the Γ-conduction band ($HH \rightarrow \Gamma c$) and from the LH band to the Γ-conduction band ($LH \rightarrow \Gamma c$). The calculated transition energies for the $HH \rightarrow \Gamma c$ and $LH \rightarrow \Gamma c$ transitions were 697 and 749 meV, respectively, which are in good agreement with the experimental results. Figure 6b shows the calculated absorption spectra of pure Ge and pseudomorphic Ge$_{0.96}$Sn$_{0.04}$ on Ge at T = 300 K. The overall absorption coefficient was the superposition of the direct interband transitions: $HH \rightarrow \Gamma c$ and $LH \rightarrow \Gamma c$; therefore, a cusp-like feature was shown in the calculated absorption spectra. The calculated result shows that the absorption coefficient decreased with an increasing wavelength.

The calculated result also shows that the optical cutoff wavelength for GeSn alloy shifted towards the longer wavelength, which considerably increased the photodetection range of the proposed device. The calculated result also shows that the absorption coefficient increased with increasing Sn concentration. For $Ge_{0.96}Sn_{0.04}$ on Ge, the absorption coefficient increased by ~3.3 times more than pure Ge at 1550 nm. This behavior can be explained based on the redshift in the absorption edges with increasing Sn concentration due to the shrinkage of the bandgap energies from the Sn alloying. Further increases in the Sn content can redshift the direct-gap absorption edge and thus extend the photodetection range of the GeSn MSM PDs for important short-wave infrared applications.

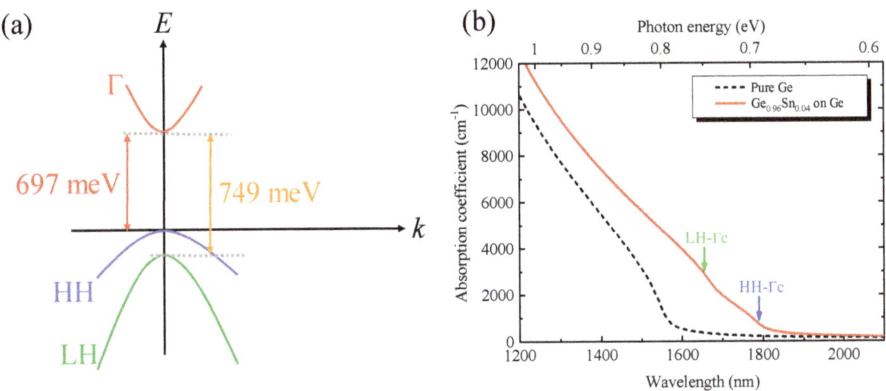

Figure 6. (**a**) Band structure of pseudomorphic GeSn with 4% Sn content on Ge in which heavy-hole (HH) and band light-hole (LH) split due to compressive strain. (**b**) Calculated absorption spectra for pure Ge and GeSn with 4% Sn composition pseudomorphically grown on Ge VS at T = 300 K.

4. Conclusions

In conclusion, we have demonstrated a GeSn MSM PD monolithically grown on Ge buffered Si substrates. The GeSn active layer was grown on a silicon substrate with good material quality as the optical absorber, which had a lower bandgap than that of pure Ge, extending the photodetection region. The responsivity experiments show enhanced spectral responsivity and low dark current compared to the existing SiGe-based MSM PDs. Furthermore, the responsivity increases with an increase in bias voltage due to the enhanced electric field. With the extended photodetection range, planar structures and CMOS compatibility, Si-based short-wave infrared GeSn PDs are promising for a wide range of applications.

Author Contributions: Conceptualization, G.-E.C.; methodology, S.G., K.-C.L., C.-H.T., Q.C, L.Z., and B.S.; formal analysis, S.G., H.K., G.-E.C., B.M., K.-C.L., C.S.T., Q.C, L.Z., M.K., and B.S.; writing—original draft preparation, S.G., Q.C.; writing—review and editing, G.-E.C., H.K., C.S.T., and B.M.; supervision, G.-E.C. All authors have read and agreed to the published version of the manuscript.

Funding: This work at CCU was funded by the Ministry of Science and Technology, Taiwan, grant numbers MOST 108-2221-E-194 -055 and MOST 109-2636-E-194 -002, and the Ministry of Education of Taiwan. The work at NTU was supported by the National Research Foundation, Singapore, under its Competitive Research Program (CRP Award NRF-CRP19-2017-01).

Acknowledgments: The authors would like to thank Yue-Tong Jheng at CCU for providing assistance in the experiments.

Conflicts of Interest: The authors declare no conflict of interest.

References

1. Vivien, L.; Rouviere, M.; Fedeli, J.-M.; Marris-Morini, D.; Damlencourt, J.-F.; Mangeney, J.; Crozat, P.; Melhaoui, L.E.; Cassan, E.; Roux, X.L.; et al. High Speed and High Responsivity Germanium Photodetector Integrated in a Silicon-On-Insulator Microwaveguide. *Opt. Express* **2007**, *15*, 9843–9848. [CrossRef]
2. Chen, L.; Lipson, M. Ultra-low Capacitance and High Speed Germanium Photodetectors on Silicon. *Opt. Express* **2009**, *17*, 7901–7906. [CrossRef]
3. Rasras, M.S.; Gill, D.M.; Earnshaw, M.P.; Doerr, C.R.; Weiner, J.S.; Bolle, C.A.; Chen, Y. CMOS Silicon Receiver Integrated with Ge Detector and Reconfigurable Optical Filter. *IEEE Photon. Tech. Lett.* **2009**, *22*, 112–114. [CrossRef]
4. Subbaraman, H.; Xu, X.; Hosseini, A.; Zhang, X.; Zhang, Y.; Kwong, D.; Chen, R.T. Recent Advances in Silicon-based Passive and Active Optical Interconnects. *Opt. Express* **2015**, *23*, 2487–2511. [CrossRef] [PubMed]
5. Vivien, L.; Polzer, A.; Marris-Morini, D.; Osmond, J.; Hartmann, J.M.; Crozat, P.; Cassan, E.; Kopp, C.; Zimmermann, H.; Fédéli, J.M. Zero-bias 40Gbit/s Germanium Waveguide Photodetector on Silicon. *Opt. Express* **2012**, *20*, 1096–1101. [CrossRef] [PubMed]
6. Wang, J.; Lee, S. Ge-photodetectors for Si-based Optoelectronic Integration. *Sensors* **2011**, *11*, 696–718. [CrossRef] [PubMed]
7. Soref, R. Silicon-based Silicon-germanium-tin Heterostructure Photonics. *Philos. Trans. R. Soc. A.* **2014**, *372*, 20130113. [CrossRef] [PubMed]
8. Deen, M.J.; Basu, P.K. *Silicon Photonics: Fundamentals and Devices*; Willey: Hoboken, NJ, USA, 2012.
9. Michel, J.; Liu, J.; Kimerlin, L.C. High-performance Ge-on-Si Photodetectors. *Nat. Photonics* **2010**, *4*, 527–534. [CrossRef]
10. Bauer, M.; Taraci, J.; Tolle, J.; Chizmeshya, A.V.G.; Zollner, S.; Smith, D.J.; Menendez, J.; Hu, C.; Kuovetakis, J. Ge-Sn Semiconductors for Band-gap and Lattice engineering. *Appl. Phys. Lett.* **2002**, *81*, 2992–2994. [CrossRef]
11. Chizmeshy, A.V.G.; Ritter, C.; Tolle, J.; Cook, C.; Menendez, J.; Kuovetakis, J. Fundamental studies of $P(GeH_3)_3$, $As(GeH_3)_3$, and $Sb(GeH_3)_3$: Practical n-dopants for New Group IV Semiconductors. *Chem. Mater.* **2006**, *18*, 6266–6277.
12. Gupta, J.P.; Bhargava, N.; Kim, S.; Adam, T.; Kolodzey, J. Infrared Electroluminescence from GeSn Heterojunction Diodes Grown by Molecular Beam Epitaxy. *Appl. Phys. Lett.* **2013**, *102*, 251117. [CrossRef]
13. Chen, J.-Z.; Li, H.; Chen, H.H.; Chang, G.-E. Structural and Optical Characteristics of $Ge_{1-x}Sn_x$/Ge Superlattices Grown on Ge-buffered Si (001) Wafers. *Opt. Mater. Express* **2014**, *4*, 1178–1185. [CrossRef]
14. Dutt, B.; Lin, H.; Sukhdeo, D.S.; Vulovic, B.M.; Gupta, S.; Nam, D.; Saraswat, K.C.; Harris, J.S., Jr. Theoretical Analysis of GeSn Alloys as a Gain Medium for a Si-Compatible Laser. *IEEE J. Sel. Top. Quantum Electron.* **2013**, *19*, 251117. [CrossRef]
15. Oehme, M.; Kostecki, K.; Arguirov, T.; Mussler, G.; Ye, K.; Golhofer, M.; Schmid, M.; Kaschel, M.; Körner, R.A.; Kittler, M.; et al. GeSn Heterojunction LEDs on Si Substrates. *IEEE Photon. Technol. Lett.* **2014**, *26*, 187–189. [CrossRef]
16. Zhou, Y.; Dou, W.; Du, W.; Pham, T.; Ghetmiri, S.A.; Al-Kabi, S.; Mosleh, A.; Alher, M.; Margetis, J.; Tolle, J.; et al. Systematic study of GeSn heterostructure-based light-emitting diodes towards mid-infrared applications. *J. Appl. Phys.* **2016**, *120*, 023102. [CrossRef]
17. Stange, D.; Driesch, N.V.D.; Rainko, D.; Roesgaard, S.; Povstugar, I.; Hartmann, J.-M.; Stoica, T.; Ikonic, Z.; Mantl, S.; Grützmacher, D.; et al. Short-wave infrared LEDs from GeSn/SiGeSn multiple quantum wells. *Optica* **2017**, *4*, 185–188. [CrossRef]
18. Sun, G.; Soref, R.A.; Cheng, H.H. Design of a Si-based lattice-matched room-temperature GeSn/GeSinSn multi-quantum-well mid-infrared laser diode. *Opt. Express* **2010**, *18*, 19957–19965. [CrossRef]
19. Chang, G.-E.; Chang, S.-W.; Chuang, S.L. Strain-Balanced Ge_zSn_{1-z}-$Si_xGe_ySn_{1-x-y}$ Multiple-Quantum-Well Lasers. *IEEE J. Quantum Electron.* **2010**, *46*, 1813–1820. [CrossRef]
20. Al-Kabi, S.; Ghetmiri, S.A.; Margetis, J.; Pham, T.; Zhou, Y.; Dou, W.; Collier, B.; Quinde, R.; Du, W.; Mosleh, A.; et al. An Optimally pumped 2.5 μm GeSn laser on Si operating at 110 K. *Appl. Phys. Lett* **2016**, *109*, 171105. [CrossRef]

21. Mukhopadhyay, B.; Sen, G.; De, S.; Basu, R.; Chakraborty, V.; Basu, P.K. Calculated Characteristics of a Transistor Laser Using Alloys of Gr-IV Elements. *Phys. Status Solidi B* **2018**, *255*, 1800117. [CrossRef]
22. Ghosh, S.; Mukhopadhyay, B.; Sen, G. Performance Enhancement of GeSn Transistor Laser with Symmetric and Asymmetric Multiple Quantum Well in the Base. *Semiconductors* **2020**, *54*, 77–84. [CrossRef]
23. Tseng, H.H.; Li, H.; Mashanov, V.; Yang, Y.J.; Cheng, H.H.; Chang, G.-E.; Soref, R.A.; Sun, G. GeSn-based p-i-n Photodiodes with Strained Active Layer on a Si wafer. *Appl. Phys. Lett.* **2013**, *103*, 231907. [CrossRef]
24. Dong, Y.; Wang, W.; Lei, D.; Gong, X.; Zhou, Q.; Lee, S.Y.; Loke, W.K.; Yoon, S.-F.; Tok, E.S.; Liang, G.; et al. Suppression of Dark Current in Germanium-tin on Silicon Surface Passivation Technique. *Opt. Express* **2015**, *23*, 18611–18619. [CrossRef] [PubMed]
25. Ghosh, S.; Mukhopadhyay, B.; Chang, G.E. Design and Analysis of GeSn-Based Resonant-Cavity-Enhanced Photodetectors for Optical Communication Applications. *IEEE Sens. J.* **2020**, *20*, 7801–7809. [CrossRef]
26. Pareek, P.; Ranjan, R.; Das, M.K. Numerical Analysis of Tin Incorporated Group IV Alloy Based MQWIP. *Opt. Quantum Electron.* **2018**, *50*, 179. [CrossRef]
27. Pareek, P.; Das, M.K.; Kumar, S. Responsivity Calculation of Group IV-based Interband MQWIP. *J. Comput. Electronics* **2017**, *17*, 319–328. [CrossRef]
28. Ghosh, S.; Mukhopadhyay, B.; Sen, G.; Basu, P.K. Performance analysis of GeSn/SiGeSn Quantum Well Infrared Photodetector in Terahertz Wavelength Region. *Phys. E* **2020**, *115*, 113692. [CrossRef]
29. Yasar, F.; Fan, W.; Ma, Z. Flexible Amorphous GeSn MSM Photodetectors. *IEEE Photon. J.* **2018**, *10*, 2800109. [CrossRef]
30. Mahmodi, H.; Hasim, M.R. Effects of Post Deposition Annealing on Crystalline State of GeSn Thin Films Sputtered on Si Substrate and Its Application to MSM Photodetector. *Mat. Res. Exp.* **2016**, *3*, 106403. [CrossRef]
31. Mahmodi, H.; Hasim, M.R.; Soga, T.; Alrokayan, S.; Khan, H.A.; Rusop, M. Synthesis of $Ge_{1-x}Sn_x$ Alloy Thin Films by Rapid Thermal Annealing of Sputtered Ge/Sn/Ge Layers on Si Substrates. *Materials* **2018**, *11*, 2248. [CrossRef]
32. Peng, Y.-H.; Cheng, H.H.; Mashanov, V.I.; Chang, G.-E. GeSn p-i-n Waveguide Photodetectors on Silicon Substrates. *Appl. Phys. Lett.* **2014**, *105*, 231109. [CrossRef]
33. Chang, G.E.; Basu, R.; Mukhopadhyay, B.; Basu, P.K. Design and Modeling of GeSn-based Heterojucntion Phototransistors for Communication Applications. *IEEE J. Sel. Quantum Electron.* **2016**, *22*, 8200409. [CrossRef]
34. Basu, R.; Chakraborty, V.; Mukhopadhyay, B.; Basu, P.K. Predicted performance of Ge/GeSn hetero-phototransistors on Si substrate at 1.55 μm. *Opt. Quantum Electron* **2015**, *47*, 387–399. [CrossRef]
35. Ghosh, S.; Mukhopadhyay, B.; Sen, G.; Basu, P.K. Study of Si-Ge-Sn based Heterobipolar Phototransistor (HPT) Exploiting Quantum Confined Stark Effect and Franz Keldysh Effect With and Without Resonant Cavity. *Phys. E* **2019**, *106*, 62–67. [CrossRef]
36. Ghosh, S.; Mukhopadhyay, B.; Sen, G.; Basu, P.K. Analysis of Some Important Parameters of Si-Ge-Sn RCE-HPT Exploiting QCSE and FKE. In Proceedings of the URSI-RCRS, Varanasi, India, 12–14 February 2020.
37. Hung, W.-T.; Barshilia, D.; Basu, R.; Cheng, H.H.; Chang, G.-E. Silicon-based High-responsivity GeSn Short-wave Infrared Heterojunction Phototransistors with a Floating Base. *Opt. Lett.* **2020**, *45*, 1088–1091. [CrossRef] [PubMed]
38. Kumar, H.; Basu, R.; Chang, G.-E. Impact of Temperature and Doping on the Performance of $Ge/Ge_{1-x}Sn_x/Ge$ Heterojunction Phototransistors. *IEEE Photonics J.* **2020**, *12*, 6801814.
39. Kumar, H.; Basu, R. Effect of Active Layer Scaling on the Performance of $Ge_{1-x}Sn_x$ Phototransistors. *IEEE Trans Electron. Dev.* **2019**, *66*, 3867–3873. [CrossRef]
40. Kumar, H.; Basu, R. Design and Analysis of $Ge/Ge_{1-x}Sn_x/Ge$ Heterojunction Phototransistor for MIR Wavelength Biological Applications. *IEEE Sensors J.* **2020**, *20*, 3504–3511. [CrossRef]
41. Sarto, A.W.; Zeghbroeck, B.J.V. Photocurrents in a Metal-Semiconductor-Metal Photodetector. *IEEE J. Quantum Electron.* **1997**, *33*, 2188–2194. [CrossRef]
42. Chuang, S.L. *Physics of Photonic Devices*, 2nd ed.; Willey: Hoboken, NJ, USA, 2012.
43. Asar, T.; SüleymanÖzçelik, Ö. Barrier enhancement of Ge MSM IR Photodetector with Ge layer Optimization. *Superlattices Microstruct.* **2015**, *88*, 685–694. [CrossRef]
44. Dushaq, G.; Ammar, N.; Mahmoud, R. Metal-Germanium-Metal Photodetector Grown on Silicon using Low Temperature RF-PECVD. *Opt. Express* **2017**, *25*, 32110–32119. [CrossRef]

45. Casalino, M.; Iodice, M.; Sirleto, L.; Rendina, I.; Coppola, G. Asymmetric MSM Sub-bandgap All-silicon Photodetector with Low Dark Current. *Opt. Express* **2013**, *21*, 28072–28082. [CrossRef] [PubMed]
46. Casalino, M.; Iodice, M.; Sirleto, L.; Rao, R.; Rendina, I.; Coppola, G. Low Dark Current Silicon-on-insulator Waveguide Metal-semiconductor-metal-photodetector Based on Internal Photoemissions at 1550 nm. *J. Appl. Phys.* **2013**, *114*, 153103. [CrossRef]
47. Zhu, S.; Lo, G.Q.; Kwong, D.L. Low-cost and High-Speed SOI Waveguide-Based Silicide Schottky-Barrier MSM Photodetectors for Broadband Optical Communications. *IEEE Photon. Tech. Lett.* **2008**, *20*, 1396–1398. [CrossRef]
48. Chang, G.-E.; Cheng, H.H. Optical Gain of Germanium Infrared Lasers on Different Crystal Orientations. *J. Phys. D Appl. Phys.* **2013**, *46*, 065103. [CrossRef]
49. Ghosh, S.; Lin, K.-C.; Tsai, C.-H.; Lee, K.H.; Chen, Q.; Son, B.; Mukhopadhyay, B.; Tan, C.S.; Chang, G.-E. Resonant-cavity-enhanced Responsivity in Germanium-on-insulator Photodetectors. *Opt. Express* **2020**, *28*, 23739–23747. [CrossRef] [PubMed]

Article

Characterization of Impact Ionization Coefficient of ZnO Based on a p-Si/i-ZnO/n-AZO Avalanche Photodiode

Gaoming Li [1,2,*], Xiaolong Zhao [1], Xiangwei Jia [1], Shuangqing Li [1] and Yongning He [1,*]

[1]   School of Microelectronics, Faculty of Electronic and Information Engineering, Xi'an Jiaotong University, No. 28 Xianning West Road, Xi'an 710049, China; zhaoxiaolong@xjtu.edu.cn (X.Z.); jixiangwei@stu.xjtu.edu.cn (X.J.); lishuangqing@stu.xjtu.edu.cn (S.L.)

[2]   School of Materials Science and Engineering, Xi'an Jiaotong University, No. 28 Xianning West Road, Xi'an 710049, China

*   Correspondence: ligaoming@xjtu.edu.cn (G.L.); yongning@mail.xjtu.edu.cn (Y.H.)

Received: 9 July 2020; Accepted: 29 July 2020; Published: 30 July 2020

Abstract: The avalanche photodiode is a highly sensitive photon detector with wide applications in optical communication and single photon detection. ZnO is a promising wide band gap material to realize a UV avalanche photodiode (APD). However, the lack of p-type doping, the strong self-compensation effect, and the scarcity of data on the ionization coefficients restrain the development and application of ZnO APD. Furthermore, ZnO APD has been seldom reported before. In this work, we employed a p-Si/i-ZnO/n-AZO structure to successfully realize electron avalanche multiplication. Based on this structure, we investigated the band structure, field profile, Current–Voltage (I-V) characteristics, and avalanche gain. To examine the influence of the width of the i-ZnO layer on the performance, we changed the i-ZnO layer thickness to 250, 500, and 750 nm. The measured breakdown voltages agree well with the corresponding threshold electric field strengths that we calculated. The agreement between the experimental data and calculated results supports our analysis. Finally, we provide data on the impact ionization coefficients of electrons for ZnO along the (001) direction, which is of great significance in designing high-performance low excess noise ZnO APD. Our work lays a foundation to realize a high-performance ZnO-based avalanche device.

Keywords: p-Si/i-ZnO/n-AZO; avalanche photodiode (APD); impact ionization coefficients

1. Introduction

The avalanche photodiode (APD) has a wide application in photoelectric conversion [1], especially in the areas of optical communication [2–5], imaging [6–8], and single photon detection [9–11], because it has a large avalanche gain that enables the high sensitivity [12,13]. When the detection spectral range enters the UV region, we need to utilize wide band gap materials as the absorption layer. Therefore, 4H-SiC [14–19], GaN [20–27]-based UV APDs have been demonstrated and investigated in recent years. ZnO is also an attractive wide band gap semiconductor that has a lot of advantages such as high exciton binding energy, direct band gap, low cost, ease of fabrication, and environmental friendliness [28–30]. Based on a simulation work, for ZnO, the ratio of the impact ionization coefficient of holes to that of electrons is rather low [31]. This has a very appealing merit that may be fully exploited to achieve high-performance, low excess noise APD devices.

Recently, some works devoted to studying the impact of ionization multiplication in ZnO/MgO [32–34] and TiO2/AlO$_x$ [35] material systems have been reported. These works are mainly based on the metal–insulator–semiconductor–insulator–metal structure. The highest electric field occurred in the insulator layer, and an avalanche gain was observed. However, this structure has two issues in our opinion. Firstly, the high electric field is primarily in the insulator area, and the field in the semiconductor layer is relatively weak, so the separation of photo carriers and the transport of these photo carriers may have a low efficiency. Secondly, the insulator has a very large band gap; then, the threshold energy needed for impact ionization is very large. Moreover, the thickness of several tens of nanometers may be too short to accelerate the initial injected carriers. So, a traditional pin structure without these issues may be more favorable. Considering the big challenge of stable and effective p-type doping for ZnO [36–38], we turned to Si for a p-type material, owing to the compatibility to the highly developed CMOS technology and the availability of a high-quality Si wafer. Although the p-Si/n-ZnO heterostructure may be exhaustively investigated [39–43], few works have been published up to date that demonstrate the impact of ionization multiplication base on this structure. The quality of the ZnO layer and the interface between Si and ZnO are the primary constraints.

In this paper, we fabricated the p-Si/i-ZnO/n-AZO structure and realized electron avalanche multiplication in the poly-crystalline i-ZnO layer. We systematically investigated this structure with an emphasis on the avalanche multiplication process. The band structure, field profile, current voltage characteristics, threshold breakdown electric field, avalanche gain, and impact ionization coefficient were comprehensively discussed. We also investigated the influence of the width of the i-ZnO layer. The major contribution from this work is that we provide the impact ionization coefficient of electrons for ZnO through triggering an electron avalanche multiplication by visible light with a wavelength of 532 nm. Our work indicates that ZnO still holds promise to realize a high-performance APD device, and it may renew the research interest in ZnO for the application of APD.

2. Materials and Methods

The p-Si/i-ZnO/n-AZO structure was fabricated through the process shown in Figure 1. At first, we cleansed the p+ Si wafer with acetone, ethanol, and deionized water successively; the whole cleansing process was with the help of ultrasonic oscillation. Then, we used Buffered Oxide Etch (BOE) to remove the native oxide layer on the surface of the wafer. Magnetron sputtering was exploited to deposit poly-crystalline ZnO and Al-doped ZnO (AZO) on the Si wafer in turn. The thickness and doping level of each layer are indicated in Figure 1. To investigate the dependence of performance on the width of the multiplication layer, which is the thickness of the i-ZnO layer, we tuned the thickness of the ZnO layer to 250, 500, and 750 nm, respectively. We also fabricated p-Si/n-AZO and p-Si/i-ZnO structures as references and to characterize the doping density based on a simpler pn junction structure. The aluminum electrodes were evaporated on AZO and Si, and the pattern was formed through photolithography. Since the AZO and Si are both heavily doped, Al electrodes can realize Ohmic contact with both layers.

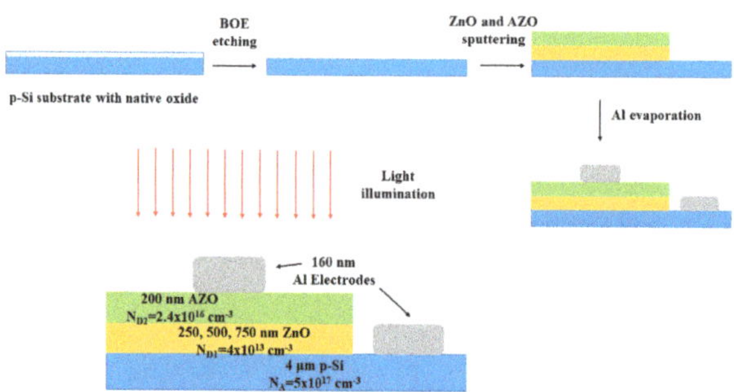

Figure 1. Fabrication process and structural details of the p-Si/i-ZnO/n-AZO heterostructure.

We utilized Capacitance–Voltage (C-V) measurement to characterize the density of doping impurities for pn junctions. To eliminate the influence of interface states as much as possible, a high frequency of 1 MHz was chosen. A source measure unit was employed to do Current–Voltage (I-V) analysis for the pin structures. To probe the avalanche gain, we used a laser diode as the light source, which provides a strong illumination with the wavelength of 532 nm. We used an attenuator to reduce the light intensity, because a relatively weak light is enough to inject some charge carriers and initiate an impact ionization multiplication process.

3. Results and Discussion

3.1. Heterostructure of Si and ZnO

At the very beginning, we want to verify that the ohmic contacts between Al electrodes and p-Si, as well as those between Al electrodes and n-AZO, were successfully formed. We fabricated two Al electrodes with a size of 10×100 µm on 500 nm-thick p-Si and 500 nm-thick n-AZO films, respectively. These p-Si and n-AZO films have an identical doping concentration to those used in the p-Si/i-ZnO/n-AZO structure. The distance between electrodes is 1 cm. The I-V curves for these two setups are shown in Figure 2, from which we can see good linearity for both I-V curves. Therefore, we confirmed that ohmic contacts were formed.

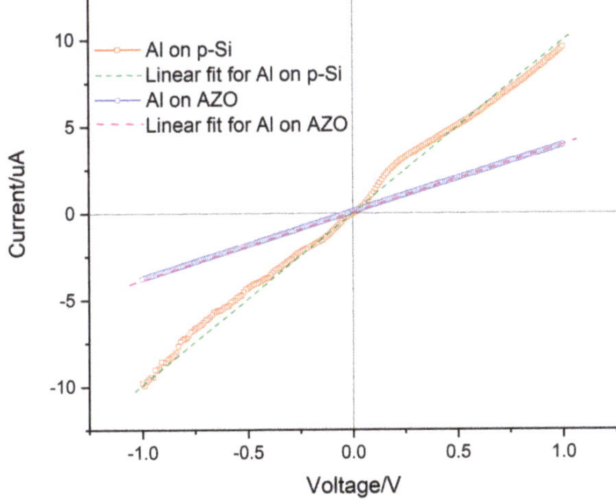

Figure 2. The Current–Voltage (I-V) curves for two Al electrodes on p-Si and n-AZO, respectively.

As a result of lacking of durable, stable, and highly effective p-type doping for ZnO, a lot of applications for ZnO based on the pn junction need another p-type material to form a heterostructure. Silicon is a very mature and widely used semiconductor material. Its planar fabrication process, for example CMOS technology, is highly developed. Compatibility with CMOS technology and the availability of high-quality Si wafer are the benefits of choosing Si as the p-type material. Besides, the lattice mismatch between Si and ZnO is tolerable. Based on the hexagonal ZnO (a = 3.252×10^{-10} m) and cubic Si (a = 5.43×10^{-10} m) lattice parameters from the literature, we can calculate the mismatch, which is 40% in this case [28]. To alleviate the strain caused by the lattice mismatch and improve the crystalline quality of ZnO near the interface, annealing at 600 degrees Celsius for 30 min in the Ar ambient was employed after the growth of ZnO film on Si. We successfully fabricated the p-Si/i-ZnO/n-AZO structure to realize the avalanche multiplication of electrons in ZnO.

The band diagram of the p-Si/i-ZnO/n-AZO structure is shown in Figure 3. The band offset, especially for the conduction band, is crucial for analyzing the electron transport. Moreover, the bending degree of the band due to the built-in potential and applied voltage bias is also responsible for the modulation of electron transport. The position of the Fermi level in the band gap is related with the doping density, which is deduced from the C-V measurement. For p-Si substrate, the doping density $N_A = 5 \times 10^{17}$ cm^{-3}, and it can be assumed that the doping impurities are all ionized at room temperature. Therefore, based on the following equation:

$$\delta_1 = E_{F,Si} - E_{v,Si} = kT \ln(N_{v,Si}/N_A) \tag{1}$$

where $E_{F,Si}$ and $E_{v,Si}$ are the Fermi level and valence band maximum (VBM) of the Si band, respectively, k is the Boltzmann constant, T is the temperature, and $N_{v,Si}$ is the effective density states in the valence band of Si, which is 1.1×10^{19} cm^{-3} [44]. Then, we can derive the energy difference between the Fermi level and VBM in Si, which is 80.4 meV.

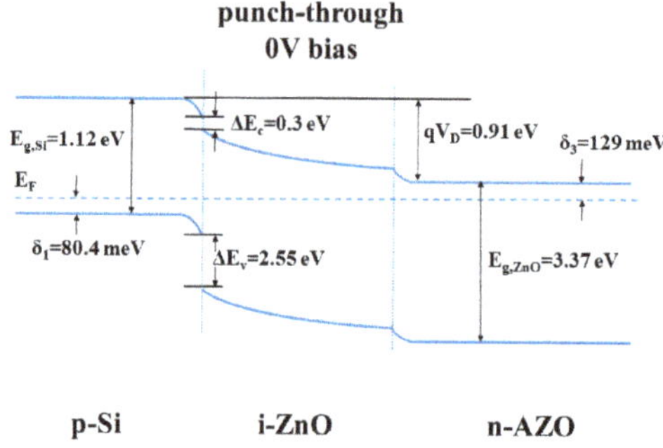

Figure 3. The band diagram of the p-Si/i-ZnO/n-AZO structure.

In a similar manner, we can deduce the energy difference between the Fermi level and conduction band minimum (CBM) in ZnO and AZO based on the following equations:

$$\delta_2 = E_{c,ZnO} - E_{F,ZnO} = kT \ln(N_{c,ZnO}/N_{D1}) \tag{2}$$

$$\delta_3 = E_{c,ZnO} - E_{F,AZO} = kT \ln(N_{c,ZnO}/N_{D2}) \tag{3}$$

where $E_{F,ZnO}$ and $E_{c,ZnO}$, $E_{F,AZO}$ and $E_{c,AZO}$ are the Fermi level and CBM for the ZnO and AZO band, respectively, $N_{c,ZnO}$ is the effective density state in the conduction band of ZnO, which is 3.4×10^{18} cm^{-3} [44], and N_{D1} and N_{D2} are the density of ionized impurities of ZnO and AZO, respectively. The values we used here were extracted from C-V measurement, which are 4×10^{13} and 2.4×10^{16} cm^{-3} for ZnO and AZO, respectively. The energy differences in the neutral regions of ZnO (δ_2) and AZO (δ_3) are 295 and 129 meV, respectively. The impurity density of unintentionally doped intrinsic ZnO is determined by C-V measurement. These impurities in ZnO are native point defects such as oxygen vacancy and interstitial zinc, which serve as donors. These native defects are the origin of the self-compensation effect and the reason why p-type doping for ZnO is very challenging.

The conduction band offset ΔE_c is determined by the difference of the electron affinity energy between Si and ZnO, which is defined as the energy measured from the bottom of the conduction band to the vacuum level. Based on the data in the literature [44,45], the electron affinity energy values for Si and ZnO are 4.05 and 4.35 eV, respectively. Therefore,

$$\Delta E_c = \chi_{ZnO} - \chi_{Si} = 0.3 \text{ eV}. \tag{4}$$

If we consider the influence of interface states, the situation could be rather complicated. Firstly, the Fermi level will be pinned by the interface states, and the bending degree of the energy band near the interface will be changed. To simplify the effect of interface, one could add a correction term to ΔE_c, which is determined by the density and position in the band gap for the interface states. Here, we did not consider the influence of interface states when calculating the band offset. It is well known that the band gaps for Si and ZnO are 1.12 and 3.37 eV, respectively. We can calculate the valence band offset ΔE_v by the equation:

$$\Delta E_v = E_{g,ZnO} - E_{g,Si} + (\chi_{ZnO} - \chi_{Si}) = 2.55 \text{ eV}. \tag{5}$$

Since the doping density of Si is about 4 orders of magnitude greater than that of ZnO, the width of the depletion layer in ZnO is much greater such that the whole region of ZnO is depleted. We will discuss the width of the depletion layer in ZnO in detail afterwards. Thus, the pin diode we fabricated belongs to the punch-through type, which has a nearly constant high field in the multiplication layer. This field profile is very suitable for avalanche multiplication.

3.2. C-V Measurement and Field Profile

The C-V measurement is a common method to acquire the doping density and built-in potential in the semiconductor field. For the ideal pn structure, we applied a reverse bias without consideration of interface states, and the capacitance of a pn junction is mainly determined by the variation of charge in the depletion layer with the applied voltage. It is easy to deduce the equation as follows:

$$\left| \frac{d1/C^2}{dV} \right| = \frac{2(\varepsilon_1 N_A + \varepsilon_2 N_D)}{q\varepsilon_1\varepsilon_2 N_D N_A} \tag{6}$$

where N_A and N_D are the doping density of the p-type and n-type layer, respectively, ε_1 and ε_2 are the relative permittivity of the n-type and p-type materials, respectively, and q is the electron charge. According to Equation (6), if we plot the $1/C^2$-V curve, we could derive the doping concentration from the slope.

To exclude the influence of interface states, a high-frequency AC field is employed. From Figures 4a and 5a, we can see that the capacitance goes down with increasing frequency, which indicates that the interface states contribute to the capacitance less at higher frequency. The reason is that the charging process through interface states is too slow to catch up with the fast varying AC field. To be more specific, the electrons and holes can be transferred between the interface states and the conduction band and valence band, respectively, depending on the position of the Fermi level. Generally speaking, the interface states that are under the Fermi level are always occupied by electrons. When the applied voltage is changing, the Fermi level will move accordingly. Then, the electric charge will flow into or out of the interface states; thus, in this case, the interfaces states behave similar to a capacitor. This charging or discharging process is usually very slow, so when the frequency of the AC voltage increases, the contribution to total capacitance from interface states gets lower. This interpretation explains the capacitance dependence on frequency well. The capacitance decreases with the increasing reverse voltage due to the increase of the width of the depletion layer. From Figures 4 and 5, based on the slope of curve that we showed using a black straight line, we derived the doping densities of i-ZnO and n-AZO, which are 4×10^{13} and 2.4×10^{16} cm^{-3}, respectively. As we can see either in Figure 4 or Figure 5, the slope of the curve for 1 MHz is not a single value, because the curve is not perfect straight line. Therefore, we took the average for the incline angles corresponding to the slope within the voltage range. The tangent of the average incline angle is the slope we extracted. The C-V measurement results are shown in Figure 6, from which we can deduce that the capacitance of the pin structure is mainly determined by the variation of the width of the depletion layer in AZO with the voltages.

When the doping concentration of each layer is determined, we can calculate the electric field profile based on Poisson's equation as follows:

$$\begin{cases} \frac{dE}{dx} = \frac{qN_A}{\varepsilon_2}, \text{ in depletion layer of Si} \\ \frac{dE}{dx} = \frac{qN_{D(1,2)}}{\varepsilon_1}, \text{ in depletion layer of ZnO or AZO} \end{cases} \tag{7}$$

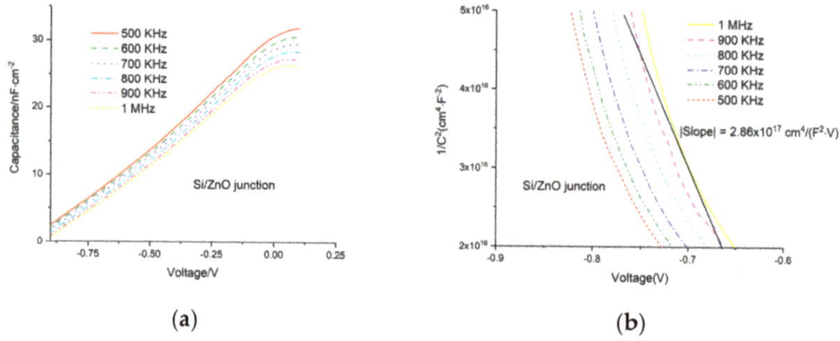

(a)

(b)

Figure 4. (**a**) Capacitance–voltage measurement data for the p-Si/i-ZnO junction; (**b**) 1/C²–V curve for p-Si/i-ZnO junction.

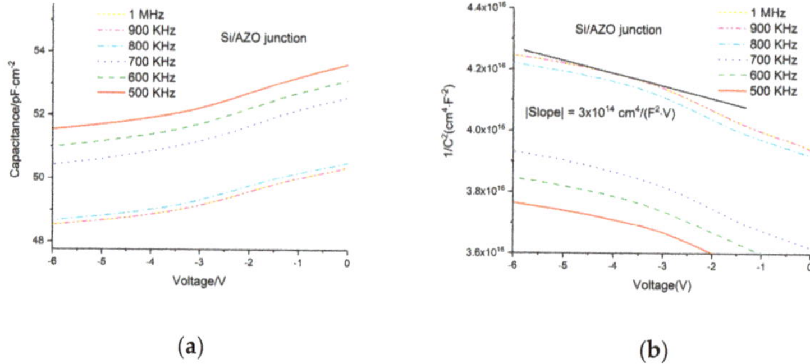

(a)

(b)

Figure 5. (**a**) Capacitance–voltage measurement data for p-Si/n-AZO junction; (**b**) 1/C²–V curve for p-Si/n-AZO junction.

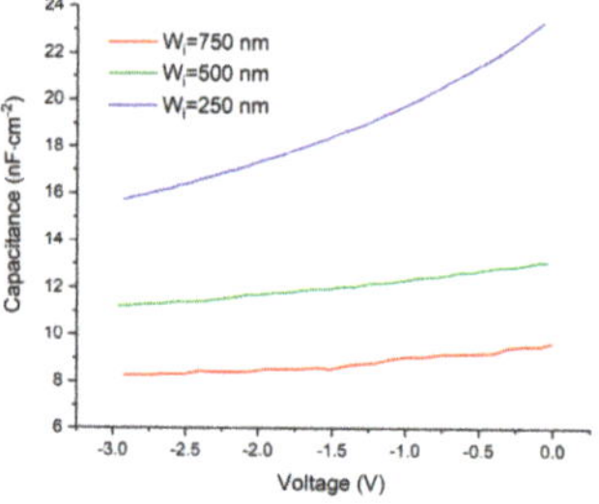

Figure 6. Capacitance–voltage measurement results measured with 1 MHz AC signal for p-Si/i-ZnO/n-AZO structure.

The calculated electric field profile for the different impurity concentrations of the i-ZnO layer is shown in Figure 7. The relative permittivities for Si (ε_2) and ZnO (ε_1) that we used in the calculation are 11.9 and 9.0, respectively [44]. The variation of doping density can alter the electric field profile. Before we do the calculation, we need to verify the assumption that the i-ZnO layer is completely depleted. It can be proved by the contradiction. If the i-ZnO layer is not fully depleted, then the width of the depletion layer should be less than the thickness of the i-ZnO layer. We can calculate the width X_D in this case by the following equation:

$$X_D = \left[\frac{2\varepsilon_{Si}\varepsilon_{ZnO}(N_A + N_{D1})^2 V_D}{qN_{D1}N_A(\varepsilon_{Si}N_A + \varepsilon_{ZnO}N_{D1})} \right]^{1/2} \approx \left[\frac{2\varepsilon_{ZnO}V_D}{qN_{D1}} \right]^{1/2}, \text{ For } N_A \gg N_{D1} \tag{8}$$

where V_D is the built-in potential for the Si/ZnO junction, which is 0.61 eV for our sample. This built-in potential was determined by the energy difference between the Fermi levels of Si and ZnO before they contact to form a heterojunction. The detailed process is outlined in the following description. From the common vacuum level, we determined the position of the CBM for Si and ZnO based on their electron affinity energy, respectively. Then, we can determine the position of the Fermi level with respect to the CBM for both of them by their impurity concentration. Finally, we calculated the difference between the Fermi levels, which is the built-in potential. Compared with the literature, this value is comparable to their results for a similar situation [46]. Based on Equation (8), we derived the width of the depletion layer, which is 1.28 µm. It is much greater than the maximum value of the i-ZnO layer, which is 750 nm. Therefore, we verified the assumption that the i-ZnO layer is fully depleted.

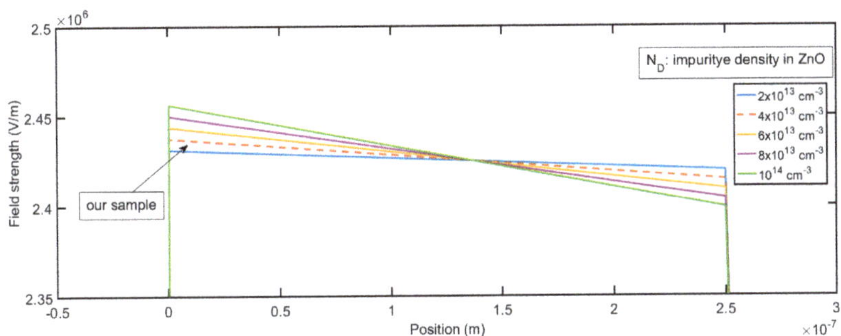

Figure 7. Field profile p-Si/i-ZnO/n-AZO structure for different impurity concentrations of i-ZnO layer.

The electric field profile in the depletion layer can be described by the following equations, which are the integration of Equation (7):

$$\begin{cases} E(x) = \frac{qN_A}{\varepsilon_{Si}}(x + W_p), \ -W_p \leq x \leq 0 \ \text{in p} - \text{Si} \\ E(x) = \frac{qN_{D1}}{\varepsilon_{ZnO}}(W_i - x) + C_1, \ 0 \leq x \leq W_i \ \text{in i} - \text{ZnO} \\ E(x) = C_1 - \frac{qN_{D2}}{\varepsilon_{ZnO}}(x - W_i), \ W_i \leq x \leq W_n \ \text{in n} - \text{AZO} \end{cases} \quad (9)$$

where W_p, W_i, and W_n are the width of the depletion layer for p-Si, i-ZnO, and n-AZO, respectively, and C_1 is a constant that can be determined by the boundary conditions. The calculated field profile for different thicknesses of the i-ZnO layer is shown in Figure 8. The greatest field strength occurs at $x = 0$, which is the interface between Si and ZnO; then, the field gradually decreases with the slope, which is proportional to the impurity density of the i-ZnO layer. If the strength of the field in the multiplication layer remains very high—in other words, it decreases very slowly—it is very suitable to realize a large gain. The smaller the impurity density of i-ZnO, the more suitable the electric field profile that we can generate. Given a constant reverse bias, the greater the width of i-ZnO, the smaller the maximum field strength that we can get. Since the doping density of Si and AZO are very high, if compared with the impurity density of ZnO, the depletion layers in Si and AZO under zero voltage bias are quite narrow. From Figures 9 and 10, we can see the variation of the depletion layer width in Si and AZO with reverse voltages. The width of the depletion layer in Si is in the range of several tens of nanometers, although the voltage reaches 40 V. In contrast, the width of the depletion layer in AZO increases faster with voltages; it can be several hundred nanometers, which is comparable to that in ZnO when the voltage goes high. However, the electric field strength in AZO falls much faster than that in ZnO, and the descending rate is determined by the impurity concentration. Therefore, the charge carriers transporting in these regions cannot initiate an impact ionization process, for they cannot gain enough energy from the electric field in such a short accelerating distance. It is apparent that a proper width of the multiplication layer is crucial to the ionization process, because if this value is too large, the field strength is not strong enough to accelerate the carrier to a critical velocity that can realize the impact ionization. On the contrary, if this value is too small, the accelerating distance would be too short to initiate an ionization multiplication. It is sound that we assume that the multiplication process solely happens in the i-ZnO when we simultaneously consider the electric field strength and acceleration distance.

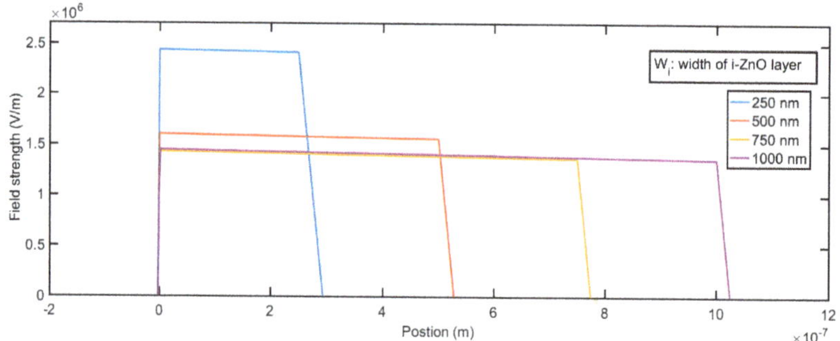

Figure 8. Field profile of the p-Si/i-ZnO/n-AZO structure for different widths of the i-ZnO layer.

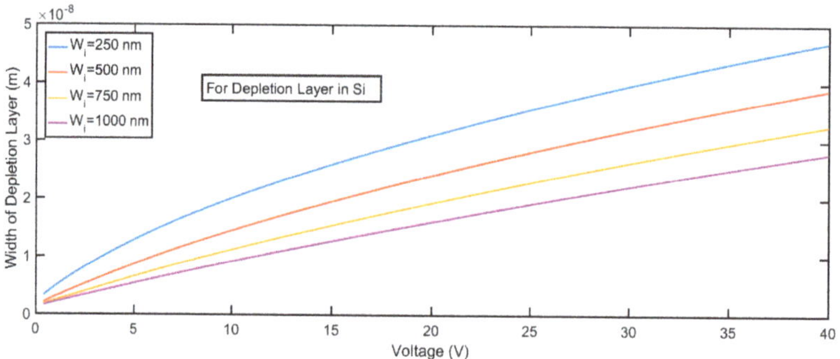

Figure 9. Width of the depletion layer in p-Si for the p-Si/i-ZnO/n-AZO structure under different voltages.

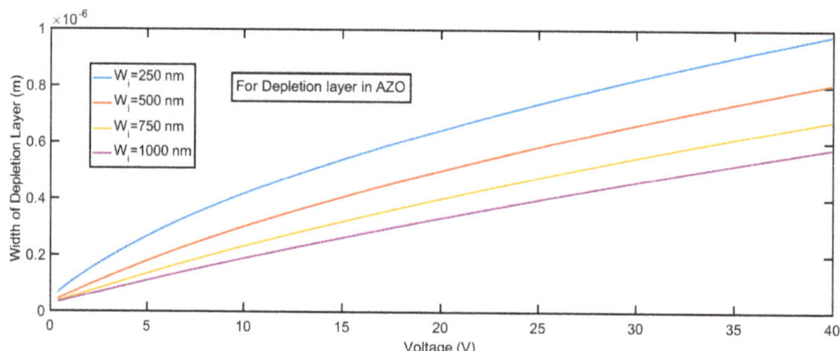

Figure 10. Width of the depletion layer in n-AZO for the p-Si/i-ZnO/n-AZO structure under different voltages.

3.3. Current–Voltage Characteristics and Influence of Width of i-ZnO Layer

The strong electric field in the depletion layer is highly needed for ionization multiplication, so APD always works under reverse voltage bias. The current–voltage curve of a p-Si/i-ZnO/n-AZO structure is shown in Figure 11. The thickness for the i-ZnO layer is 250 nm. It has typical I-V characteristics for an avalanche photodiode (APD). The reverse voltage range can be broken into three regions. Within the first region, APD is in a simple diode mode without a gain or with a unit gain. In this region, the current is equal to the reverse saturation current. In the second region, APD works in a linear mode, in which the photocurrent is proportional to the light intensity, and the photocurrent decays rapidly after the light is off. The third region is related to the Geiger mode, in which an avalanche multiplication process is generated violently such that the generation rate of carriers is greater than the collection rate of the electrodes. So, we need a quenching mechanism to cease this multiplication process after the light is off. The boundary between the linear mode and Geiger mode is the critical voltage, which is called the avalanche breakdown voltage.

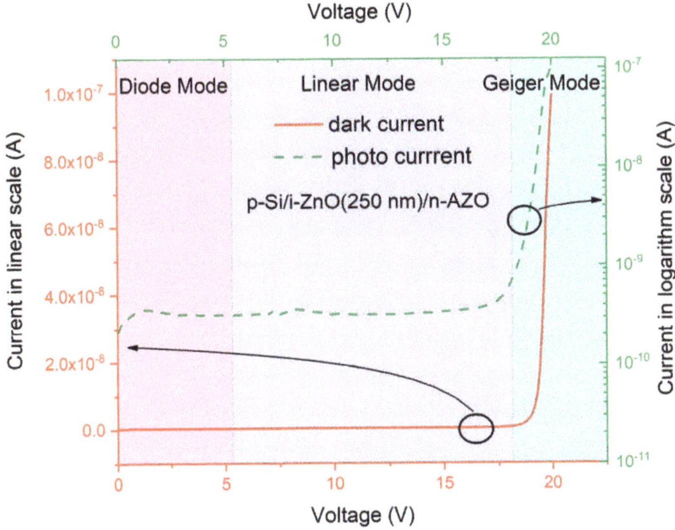

Figure 11. The typical I-V characteristics for our p-Si/i-ZnO/n-AZO avalanche photodiode; the green dashed line indicates the photo current, and the red solid line represents the dark current.

It is easy to deduce that the threshold energy E_i demanded for impact ionization is highly related to the band structure of the semiconductor, which serves as a multiplication layer. Regardless of whether there are two parabolic bands, three parabolic bands, or non-parabolic bands, E_i is comparable with the band gap E_g. As the well-known 3/2-band-gap rule indicates, if the effective mass for electrons and holes are equal, the threshold energy E_i is equal to $3E_g/2$ [12]. Since the effective mass of the holes in ZnO is lacking, we took the assumption that the effective mass of the holes is equal to that of the electrons. Then, if we make a sound guess for the mean free path of the carrier, which is defined as the distance that the carrier moves between two collisions with phonons, we could roughly derive the threshold electric field strength required for impact ionization multiplication. The mean free path can be expressed as the product of the carrier velocity and the scattering time. The scattering time is the reciprocal of the scattering rate. Based on the aforementioned analysis, the threshold field strength can be described as follows:

$$E_{BR} = \frac{(3m^*E_g)^{1/2} \cdot R(3E_g/2)}{q} \tag{10}$$

where m^* is the effective mass of the charge carrier, which is equal to 0.27 m_0 for ZnO [44]. m_0 is the electron rest mass. q is the electron charge. $R(3E_g/2)$ is the scattering rate for the carrier with the energy, which is equal to $3E_g/2$. We substitute the quantities in Equation (10) with realistic values; then, we acquire that E_{BR} is equal to 4.2×10^5 V/cm. The R($3E_g/2$) we used is 1.1×10^{13} s^{-1} from the reference [31].

We calculated the maximum electric field strength at various reverse voltages at the interface between ZnO and Si for different i-ZnO layer thicknesses. The calculated results are shown in Figure 12. As we can see, given the same reverse voltage applied, the thicker the i-ZnO layer, the smaller the maximum electric field we can get. The reverse voltages corresponding to E_{BR} are avalanche breakdown voltages; for the i-ZnO layer with different thicknesses, they are 25 V (250 nm), 35 V (500 nm), and 45 V (750 nm), respectively. Compared with the I-V curves for the p-Si/i-ZnO/n-AZO APD, the experimental results are smaller than the calculated results. Possible reasons for this discrepancy could be that the scattering rate we used is greater than that in the realistic situation, and the assumption and approximation may bring some errors.

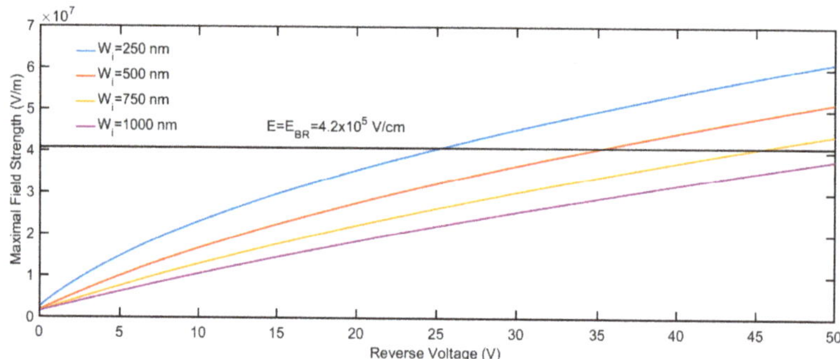

Figure 12. The maximum electric field strength at various reverse voltages.

3.4. Avalanche Gain and Ionization Coefficient

We utilized a laser diode with wavelength of 532 nm as the light source to investigate the photo response and avalanche gain of our p-Si/i-ZnO/n-AZO APD. Although the light firstly shines on the AZO side, which is on the top of the device, photocarriers will be generated in the p-Si, which is the bottom layer of the device. The reason is that the wavelength of 532 nm is much longer than the cutoff wavelength of ZnO and AZO; thus, this green light will pass through the ZnO and AZO layer without much loss. The photocarrier injection happens in the p-Si layer, and the minority carrier electrons will drift to the ZnO layer under the strong built-in electric field. So, these photogenerated electrons are the initial electrons that will be accelerated in a high-field ZnO layer and triggered the impact ionization multiplication. Based on this analysis, we can draw two important conclusions: first, the impact ionization multiplication happens in the i-ZnO layer; second, this avalanche multiplication is an electron-dominant process, and the contribution from holes can be neglected. Moreover, from the calculated results in the literature [31], we know that for ZnO, the ionization coefficient of the electrons (α) is much greater than that of the holes (β). This also supports the conclusion that we can omit the influence of the holes and solely consider the electrons' impact ionization.

The I-V curves of the APD under illumination are shown in Figure 13. The avalanche gain can represent the degree to which the photocurrent is magnified through impact ionization multiplication. The definition is based on the following equation:

$$M = \frac{I_{MP} - I_{MD}}{I_P - I_D} \tag{11}$$

where I_P and I_D are the photocurrent and dark current without multiplication, respectively, and I_{MP} and I_{MD} are the photocurrent and dark current with avalanche multiplication. It is apparent that the ionization multiplication is highly dependent on the strength of electric field; thus, the avalanche gain is a function of reverse voltage. We calculated the avalanche gain using Equation (11). The result is shown in Figure 14. When we calculated the gain, we assumed that the gain is a unit at 5 V and used the photocurrent and dark current at 5 V for the denominator of the right part of Equation (11).

From the relationship between the avalanche gain and the electric field strength, we can estimate that the breakdown voltage is about 3.3×10^5 V/cm.

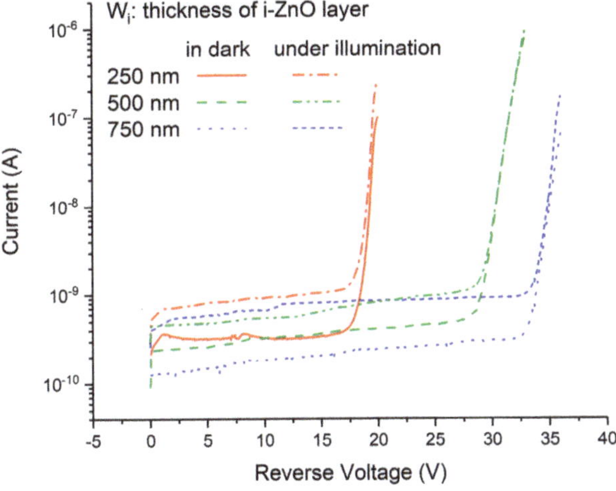

Figure 13. The I-V characteristics of p-Si/i-ZnO/n-AZO APD in dark and under illumination.

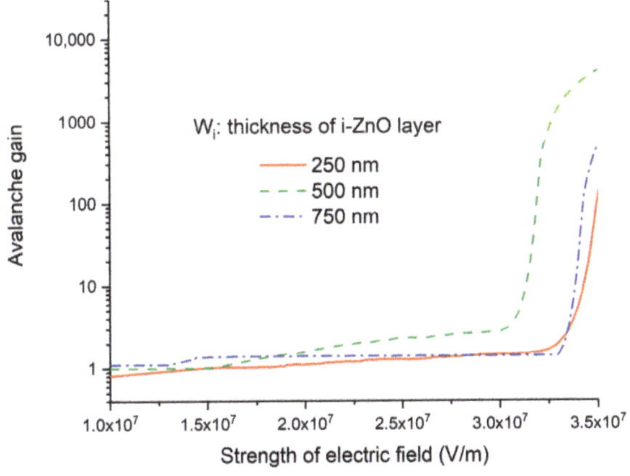

Figure 14. The avalanche gain at various reverse voltages.

Ionization coefficients are crucial parameters that mean how many electrons and holes are created when the initial carrier travels along the electric field for a unit distance. They are related with the performance of APD such as the multiplication gain, excess noise, threshold energy, and gain–bandwidth product. When one designs a high-performance APD, the ionization coefficient is an important factor that should always be kept in mind. In a pin structure, the ionization coefficients are independent of position; therefore, the avalanche gain can be expressed using ionization coefficients as the following equation shows:

$$M = \frac{(1 - \beta/\alpha)\exp[\alpha W_D(1 - \beta/\alpha)]}{1 - (\beta/\alpha)\exp[\alpha W_D(1 - \beta/\alpha)]} \approx \exp(\alpha W_D), \text{ if } \alpha \gg \beta \tag{12}$$

where W_D is the width of the multiplication layer. According to Equation (12), we calculated the ionization coefficient of electron (α) for APD with i-ZnO layers with different thicknesses. The calculation process is as follows. We firstly calculated the avalanche gain under various voltages;

then, based on the relationship between the ionization coefficient and avalanche gain, we derived the ionization coefficient under these voltages. In an ideal situation, the avalanche gain and ionization coefficient should have been independent of the specific structural parameters of the punch-through pin structure. They are mainly determined by what material is utilized as a multiplication layer. However, owing to the nonuniformity of the ZnO layer, the avalanche gain and ionization coefficient under a given electric field extracted from the I-V curves exhibit a little discrepancy for different ZnO layer thicknesses. Therefore, we averaged the ionization coefficients derived from the pin structure with different ZnO layer thicknesses to minimize the extraction error. The results are shown in Figure 15. It should be noted that these ionization coefficients correspond to the case in which the electric field is along the (001) direction of ZnO, because our ZnO sample is highly c-axis oriented and the electric field is along the c-axis, which is the (001) direction. The results represented herein show that p-Si/i-ZnO/n-AZO is a promising structure to realize high-performance APD.

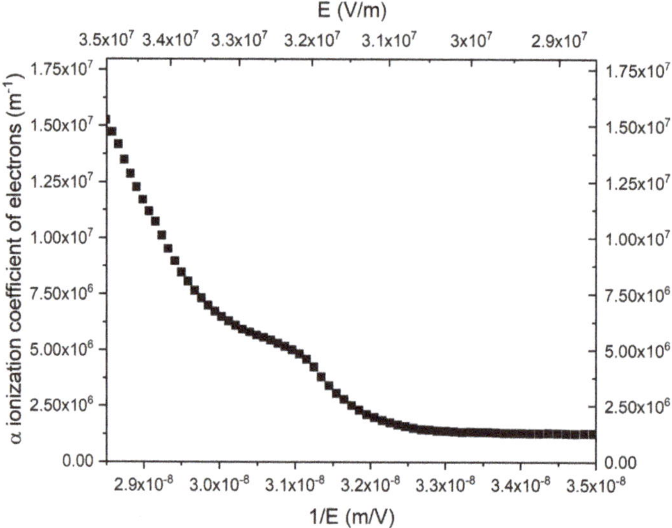

Figure 15. The impact ionization coefficients of electrons.

4. Conclusions

In this paper, we have fabricated a p-Si/i-ZnO/n-AZO structure and realized avalanche multiplication in the poly-crystalline i-ZnO layer based on this structure. The band structure of this p-Si/i-ZnO/n-AZO heterostructure was systematically investigated. Based on C-V measurement, the ionized impurity density of the unintentionally doped ZnO layer is 4×10^{13} cm^{-3}, while the doping density of the AZO layer is 2.4×10^{16} cm^{-3}. This i-layer with rather low impurity density is very important for a suitable field profile and ionization multiplication. We calculated the field profile for our p-Si/i-ZnO/n-AZO structure and investigated how the impurity density of the i-ZnO layer and the width of the i-ZnO layer affect the field profile. We chose three values, which are 250, 500, and 750 nm for the width of the i-ZnO layer. The calculated results show that the p-Si/i-ZnO/n-AZO structure of the parameters mentioned before is a punch-through type pin APD, which holds a very high field in the i layer. According to the field profile and width of the depletion layer in AZO and Si, an avalanche multiplication process only happened in the i-ZnO layer. We proposed an equation that roughly estimates the threshold electric field strength for the avalanche breakdown and breakdown voltage for different i-ZnO layer widths. The threshold electric field strength for avalanche breakdown for our pin structure is 3.3×10^5 V/cm. To initiate the electron dominant avalanche ionization multiplication process in the i-ZnO layer and derive the ionization coefficient of electrons, we used 532 nm light from a laser diode to illuminate the pin APD. The photocurrent is greatly enhanced when the ionization multiplication is triggered at high voltages. A large avalanche gain was observed, and its variation with the electric field strength was investigated. Then, we derived the impact ionization coefficient of electrons for ZnO along the (001) direction, which is a very significant parameter for APD design and optimization. Our work provides important data for high-performance APD based on an Si/ZnO heterostructure, and it also indicates that the p-Si/i-ZnO/n-AZO structure is a promising option to realize a high-performance avalanche-type device. It may renew the research interest in the realization of APD using ZnO.

Author Contributions: Conceptualization, G.L., X.Z.; methodology, X.Z., X.J., S.L.; validation, G.L., X.J. and S.L.; formal analysis, G.L.; investigation, G.L., X.J. and S.L.; resources, Y.H.; data curation, X.J., S.L.; writing—original draft preparation, G.L.; supervision, G.L.; project administration, G.L., Y.H. All authors have read and agreed to the published version of the manuscript.

Funding: This research was funded by the Youth Program of National Science Foundation of China, grant muber 61804121; General Program of National Natural Science Foundation of China, grant number 61574113; China Postdoctoral Science Foundation, grant number 2018M643650; Research Support Plan for New Teachers of Xi'an Jiaotong University.

Acknowledgments: The authors would like to thank Shuwen Guo for her assistance in C-V measurement.

Conflicts of Interest: The authors declare no conflict of interest. The funders had no role in the design of the study; in the collection, analyses, or interpretation of data; in the writing of the manuscript, or in the decision to publish the results.

References

1. Campbell, J.C. Recent Advances in Avalanche Photodiodes. *J. Lightwave Technol.* **2016**, *34*, 278–285. [CrossRef]
2. Campbell, J.C. Advances in photodetectors. In *Optical Fiber Telecomunications, Part A: Components and Subsystems*, 5th ed.; Kaminow, I., Li, T., Wilner, A.E., Eds.; Academic: San Francisco, CA, USA, 2008; Volume 5.
3. Xu, Z.Y.; Sadler, B.M. Ultraviolet communications: Potential and state-of-the-art. *IEEE Commun. Mag.* **2008**, *46*, 67–73.
4. Assefa, S.; Xia, F.N.A.; Vlasov, Y.A. Reinventing germanium avalanche photodetector for nanophotonic on-chip optical interconnects. *Nature* **2010**, *464*, 80–84. [CrossRef]
5. Kato, K. Ultrawide-band/high-frequency photodetectors. *IEEE Trans. Microw. Theory Tech.* **1999**, *47*, 1265–1281. [CrossRef]
6. Bertone, N.; Clark, W. Avalanche photodiode arrays provide versatility in ultrasensitive applications. *Laser Focus World* **2007**, *43*, 69–72.
7. Renker, D.; Lorenz, E. Advances in solid state photon detectors. *J. Instrum.* **2009**, *4*, P04004. [CrossRef]
8. Woodson, M.E.; Ren, M.; Maddox, S.J.; Chen, Y.J.; Bank, S.R.; Campbell, J.C. Low-noise AlInAsSb avalanche photodiode. *Appl. Phys. Lett.* **2016**, *108*, 081102. [CrossRef]
9. Hayden, O.; Agarwal, R.; Lieber, C.M. Nanoscale avalanche photodiodes for highly sensitive and spatially resolved photon detection. *Nat. Mater.* **2006**, *5*, 352–356. [CrossRef]
10. Faramarzpour, N.; Deen, M.J.; Shirani, S.; Fang, Q. Fully integrated single photon avalanche diode detector in standard CMOS 0.18-mu m technology. *IEEE Trans. Electron. Devices* **2008**, *55*, 760–767. [CrossRef]
11. Lacaita, A.; Francese, P.A.; Zappa, F.; Cova, S. Single-Photon Detection Beyond 1-Mu-M-Performance of Commercially Available Germanium Photodiodes. *Appl. Optics* **1994**, *33*, 6902–6918. [CrossRef]
12. Capasso, F. Physics of avalanche photodiodes. In *Lightwave Communications Techonolgy Part D, Photodetectors*; Tsang, W.T., Ed.; Academic Press: Orlando, FL, USA, 1985; Volume 22.
13. Stillman, G.E.; Wolfe, C.M. Avalanche photodiodes. In *Infrared Detectors II*; Willardson, R.K., Beer, A.C., Eds.; Academic Press: New York, NY, USA, 1977; Volume 12.
14. Su, L.L.; Zhou, D.; Lu, H.; Zhang, R.; Zheng, Y.D. Recent progress of SiC UV single photon counting avalanche photodiodes. *J. Semicond.* **2019**, *40*, 121802. [CrossRef]
15. Hu, J.; Xin, X.B.; Li, X.Q.; Zhao, J.H.; VanMil, B.L.; Lew, K.K.; Myers-Ward, R.L.; Eddy, C.R.; Gaskill, D.K. 4H-SiC visible-blind single-photon avalanche diode for ultraviolet detection at 280 and 350 nm. *IEEE Trans. Electron. Devices* **2008**, *55*, 1977–1983. [CrossRef]

16. Chen, X.P.; Zhu, H.L.; Cai, J.F.; Wu, Z.Y. High-performance 4H-SiC-based ultraviolet p-i-n photodetector. *J. Appl. Phys.* **2007**, *102*, 024505. [CrossRef]
17. Cai, X.L.; Zhou, D.; Yang, S.; Lu, H.; Chen, D.J.; Ren, F.F.; Zhang, R.; Zheng, Y.D. 4H-SiC SACM Avalanche Photodiode with Low Breakdown Voltage and High UV Detection Efficiency. *IEEE Photonics J.* **2016**, *8*, 6805107. [CrossRef]
18. Bai, X.G.; Guo, X.Y.; Mcintosh, D.C.; Liu, H.D.; Campbell, J.C. High detection sensitivity of ultraviolet 4H-SiC avalanche photodiodes. *IEEE J. Quantum Electron.* **2007**, *43*, 1159–1162. [CrossRef]
19. Zhou, X.Y.; Tan, X.; Lv, Y.J.; Wang, Y.G.; Li, J.; Han, T.T.; Guo, H.Y.; Liang, S.X.; Zhang, Z.H.; Feng, Z.H.; et al. 8 × 8 4H-SiC Ultraviolet Avalanche Photodiode Arrays with High Uniformity. *IEEE Electron. Device Lett.* **2019**, *40*, 1591–1594. [CrossRef]
20. Reddy, P.; Breckenridge, M.H.; Guo, Q.; Klump, A.; Khachariya, D.; Pavlidis, S.; Mecouch, W.; Mita, S.; Moody, B.; Tweedie, J.; et al. High gain, large area, and solar blind avalanche photodiodes based on Al-rich AlGaN grown on AlN substrates. *Appl. Phys. Lett.* **2020**, *116*, 081101. [CrossRef]
21. Ji, D.; Ercan, B.; Benson, G.; Newaz, A.K.M.; Chowdhury, S. 60 A/W high voltage GaN avalanche photodiode demonstrating robust avalanche and high gain up to 525K. *Appl. Phys. Lett.* **2020**, *116*, 211102. [CrossRef]
22. Nikzad, S.; Hoenk, M.; Jewell, A.D.; Hennessy, J.J.; Carver, A.G.; Jones, T.J.; Goodsall, T.M.; Hamden, E.T.; Suvarna, P.; Bulmer, J.; et al. Single Photon Counting UV Solar-Blind Detectors Using Silicon and III-Nitride Materials. *Sensors* **2016**, *16*, 927. [CrossRef]
23. Verghese, S.; McIntosh, K.A.; Molnar, R.J.; Mahoney, L.J.; Aggarwal, R.L.; Geis, M.W.; Molvar, K.M.; Duerr, E.K.; Melngailis, I. GaN avalanche photodiodes operating in linear-gain mode and Geiger mode. *IEEE Trans. Electron. Devices* **2001**, *48*, 502–511. [CrossRef]
24. Carrano, J.C.; Lambert, D.J.H.; Eiting, C.J.; Collins, C.J.; Li, T.; Wang, S.; Yang, B.; Beck, A.L.; Dupuis, R.D.; Campbell, J.C. GaN avalanche photodiodes. *Appl. Phys. Lett.* **2000**, *76*, 924–926. [CrossRef]
25. Yang, B.; Li, T.; Heng, K.; Collins, C.; Wang, S.; Carrano, J.C.; Dupuis, R.D.; Campbell, J.C.; Schurman, M.J.; Ferguson, I.T. Low dark current GaN avalanche photodiodes. *IEEE J. Quantum Electron.* **2000**, *36*, 1389–1391. [CrossRef]
26. Zhou, Q.G.; McIntosh, D.C.; Lu, Z.W.; Campbell, J.C.; Sampath, A.V.; Shen, H.E.; Wraback, M. GaN/SiC avalanche photodiodes. *Appl. Phys. Lett.* **2011**, *99*, 131110. [CrossRef]
27. McClintock, R.; Razeghi, M. Ultraviolet avalanche photodiodes. In *Optical Sensing, Imaging, And Photon Counting: Nanostructured Devices and Applications, Proceedings of SPIE*; SPIE: Bellingham, DC, USA, 2015; Volume 9555, p. 95550B.
28. Ozgur, U.; Alivov, Y.I.; Liu, C.; Teke, A.; Reshchikov, M.A.; Dogan, S.; Avrutin, V.; Cho, S.J.; Morkoc, H. A comprehensive review of ZnO materials and devices. *J. Appl. Phys.* **2005**, *98*, 041301. [CrossRef]
29. Look, D.C. Recent advances in ZnO materials and devices. *Mater. Sci. Eng. B-Solid State Mater. Adv. Technol.* **2001**, *80*, 383–387. [CrossRef]
30. Janotti, A.; Van de Walle, C.G. Fundamentals of zinc oxide as a semiconductor. *Rep. Prog. Phys.* **2009**, *72*, 126501. [CrossRef]
31. Bertazzi, F.; Penna, M.; Goano, M.; Bellotti, E. Theory of high field carrier transport and impact ionization in ZnO. In *Conference of Oxide-based Materials and Devices, Proceedings of SPIE*; SPIE: Bellingham, DC, USA, 2010; Volume 7603, p. 760303.
32. Yu, J.; Shan, C.X.; Huang, X.M.; Zhang, X.W.; Wang, S.P.; Shen, D.Z. ZnO-based ultraviolet avalanche photodetectors. *J. Phys. D Appl. Phys.* **2013**, *46*, 305105. [CrossRef]
33. Yu, J.; Shan, C.X.; Qiao, Q.; Xie, X.H.; Wang, S.P.; Zhang, Z.Z.; Shen, D.Z. Enhanced Responsivity of Photodetectors Realized via Impact Ionization. *Sensors* **2012**, *12*, 1280–1287. [CrossRef]
34. Zhu, H.; Shan, C.X.; Wang, L.K.; Zheng, J.; Zhang, J.Y.; Yao, B.; Shen, D.Z. Metal-Oxide-Semiconductor-Structured MgZnO Ultraviolet Photodetector with High Internal Gain. *J. Phys. Chem. C* **2010**, *114*, 7169–7172. [CrossRef]
35. Wang, W.J.; Shan, C.X.; Zhu, H.; Ma, F.Y.; Shen, D.Z.; Fan, X.W.; Choy, K.L. Metal-insulator-semiconductor-insulator-metal structured titanium dioxide ultraviolet photodetector. *J. Phys. D Appl. Phys.* **2010**, *43*, 045102. [CrossRef]
36. Zhang, S.B.; Wei, S.H.; Zunger, A. Intrinsic n-type versus p-type doping asymmetry and the defect physics of ZnO. *Phys. Rev. B* **2001**, *63*, 075205. [CrossRef]

37. Fan, J.C.; Sreekanth, K.M.; Xie, Z.; Chang, S.L.; Rao, K.V. *p*-Type ZnO materials: Theory, growth, properties and devices. *Prog. Mater. Sci.* **2013**, *58*, 874–985. [CrossRef]

38. Look, D.C.; Claftin, B. P-type doping and devices based on ZnO. *Phys. Status Solidi B* **2004**, *241*, 624–630. [CrossRef]

39. Narayan, J.; Sharma, A.K.; Kvit, A.; Jin, C.; Muth, J.F.; Holland, O.W. Novel cubic $Zn_xMg_{1-x}O$ epitaxial hetero structures on Si (100) substrates. *Solid State Commun.* **2002**, *121*, 9–13. [CrossRef]

40. Shen, L.; Ma, Z.Q.; Shen, C.; Li, F.; He, B.; Xu, F. Studies on fabrication and characterization of a ZnO/p-Si-based solar cell. *Superlattices Microstruct.* **2010**, *48*, 426–433. [CrossRef]

41. Chand, S.; Kumar, R. Electrical characterization of Ni/n-ZnO/p-Si/Al heterostructure fabricated by pulsed laser deposition technique. *J. Alloy. Compd.* **2014**, *613*, 395–400. [CrossRef]

42. Zhang, X.M.; Golberg, D.; Bando, Y.; Fukata, N. n-ZnO/p-Si 3D heterojunction solar cells in Si holey arrays. *Nanoscale* **2012**, *4*, 737–741. [CrossRef]

43. Li, X.P.; Zhang, B.L.; Zhu, H.C.; Dong, X.; Xia, X.C.; Cui, Y.G.; Ma, Y.; Du, G.T. Study on the luminescence properties of n-ZnO/p-Si hetero-junction diode grown by MOCVD. *J. Phys. D Appl. Phys.* **2008**, *41*, 035101. [CrossRef]

44. Sze, S.M.; Ng, K.K. *Physics of Semiconductor Devices*, 3rd ed.; John Wiley & Sons: Hoboken, NJ, USA, 2007.

45. Sun, H.; Zhang, Q.F.; Wu, J.L. Electroluminescence from ZnO nanorods with an n-ZnO/p-Si heterojunction structure. *Nanotechnology* **2006**, *17*, 2271–2274. [CrossRef]

46. Romeo, R.; Lopez, M.C.; Leinen, D.; Martin, F.; Ramos-Barrado, J.R. Electrical properties of the n-ZnO/c-Si heterojunction prepared by chemical spray pyrolysis. *Mater. Sci. Eng. B* **2004**, *110*, 87–93. [CrossRef]

 *micromachines*

Article

Theoretical Investigation of Near-Infrared Fabry–Pérot Microcavity Graphene/Silicon Schottky Photodetectors Based on Double Silicon on Insulator Substrates

Maurizio Casalino

Institute of Applied Science and Intelligent Systems "Eduardo Caianiello" (CNR), Via P. Castellino n. 141, 80131 Naples, Italy; maurizio.casalino@na.isasi.cnr.it

Received: 29 June 2020; Accepted: 20 July 2020; Published: 22 July 2020

Abstract: In this work a new concept of silicon resonant cavity enhanced photodetector working at 1550 nm has been theoretically investigated. The absorption mechanism is based on the internal photoemission effect through a graphene/silicon Schottky junction incorporated into a silicon-based Fabry–Pérot optical microcavity whose input mirror is constituted by a double silicon-on-insulator substrate. As output mirror we have investigated two options: a distributed Bragg reflector constituted by some periods of silicon nitride/hydrogenated amorphous silicon and a metallic gold reflector. In addition, we have investigated and compared two configurations: one where the current is collected in the transverse direction with respect to the direction of the incident light, the other where it is collected in the longitudinal direction. We show that while the former configuration is characterized by a better responsivity, spectral selectivity and noise equivalent power, the latter configuration is superior in terms of bandwidth and responsivity × bandwidth product. Our results show responsivity of 0.24 A/W, bandwidth in GHz regime, noise equivalent power of 0.6 nW/cm$\sqrt{\text{Hz}}$ and full with at half maximum of 8.5 nm. The whole structure has been designed to be compatible with silicon technology.

Keywords: resonant cavity; photodetectors; near-infrared; silicon; graphene

---

1. Introduction

Silicon (Si) photonics is nowadays an emerging market promising to reach a value of $560 M at chip level and $4 B at transceiver level in 2025 as shown in Figure 1. Indeed, both switching and interconnects of the existing data center risk becoming an early bottleneck for the huge increase in internet data traffic driven by social network and video contents.

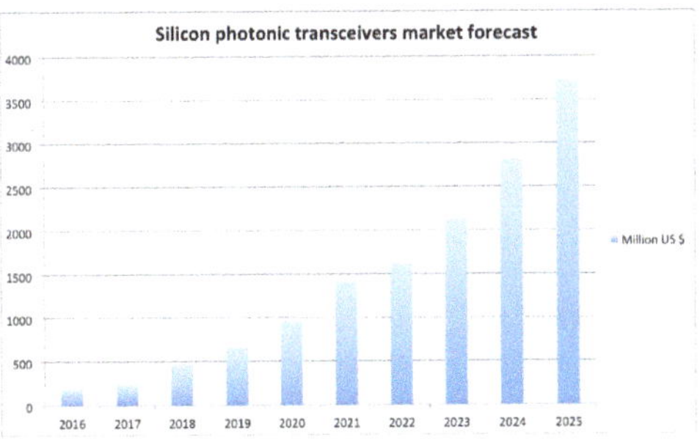

Figure 1. Silicon photonics 2016–2025 market forecast [1].

This is because in the future new technologies must necessarily be introduced for making Si fully compatible for sensors [2,3] and photonic devices in general. Since the 1980s the interest of both scientists and industry in Si photonics has grown exponentially. Nowadays Luxtera together with Intel share the leadership in Si photonics with commercial transceivers able to transmit data at a rate of 100 G. Si is a very mature technology and still plays a key role in the microelectronic industry, for this reason the realization of photonic devices in Si would be the best approach for matching the data center requirements in terms of reliability, low cost, power consumption and integration density.

Photodetectors (PDs) are key devices in photonics making possible the transduction of light into current. Si PDs operating in the visible spectrum are still commercial components, on the other hand Si employment for near-infrared (NIR) detection is hindered by its optical transparence over 1.1 micron.

Currently, Si-based NIR PDs take advantage of the integration of germanium (Ge) [4,5], however, these devices are characterized by high leakage current due to the lattice mismatch of 4.3%. In order to mitigate this drawback a buffer layer, gradually matching the Si to the Ge lattice, can be fabricated [5–7]; even if this approach reduces the leakage current, however it remains quite high. In addition, the fabrication of this buffer layer involves high thermal budget fabrication processes [8] which prevent Ge being monolithically integrated on Si. Finally, the low Ge absorption at 1550 nm (one order of magnitude lower than indium gallium arsenide, InGaAs) hinders the realization of high-speed PIN devices due to the high thickness of the intrinsic region.

Internal photoemission effect (IPE) offers one option to the all-Si approach in the field of the NIR detection. IPE concerns the optical absorption of a metal involved in a Schottky junction and then the emission of the photo-excited carriers into the semiconductor over the Schottky junction [9,10]. Both palladium silicide (Pd_2Si) and platinum silicide (PtSi) have been widely employed in infrared charge-coupled device (CCD) image sensors but, unfortunately, they have to work at cryogenic temperature for increasing the signal-to-noise ratio at an acceptable level. Pd_2Si/Si Schottky PDs can operate in the spectrum ranging from 1 to 2.4 µm requiring temperatures of 120 K [11,12] (e.g., satellite applications) while PtSi/Si Shottky PDs can work to an extended spectrum ranging from 3 to 5 µm [13,14] but they needs to operate at lower temperatures of only 80 K. Focal plane arrays (FPA) based on 512 × 512 PtSi/Si PDs have been demonstrated [15].

In 2006, for first time, it was theoretically proposed to use IPE for detecting NIR in Si at room temperature. The idea was to work with junctions characterized by higher Schottky barriers for reducing the dark current and, at the same time, to recovery the unavoidable reduced efficiency by incorporating the junction into a Fabry–Peròt optical microcavity [16]. After that, many innovative approaches have been investigated taking advantage of: Si nanoparticles (NPs) [17], surface plasmon polaritons (SPPs) [18,19], antennas [20] and gratings [21]. Despite this, to date responsivities of only 30 mA/W [17] and 5 mA/W [22] were reported for waveguide and free-space PDs, respectively. The low responsivity is mainly due to the low emission probability of the charge carriers excited from the metal to the semiconductor. IPE theory shows that this emission probability can be increased by thinning the metal [23,24]. This is the reason why the idea of replacing metal with graphene was born. Graphene/Si Schottky PDs have shown unexpected efficiency in both the visible [25,26] and NIR [27] spectrum: while in the visible range this enhancement has been ascribed to the gating effect of the graphene/SiO_2/Si capacitor in parallel to the graphene/Si junction [25,28], in the NIR range increased IPE has been attributed to the increased charge emission probability due to the mediation of the interface defects [27]. However, in this last case the whole efficiency is hindered by the low graphene absorption (only 2.3%). In order to increase the graphene optical absorption, many strategies have been followed: by realizing plasmonic nanostructures [29], by reducing the graphene size down to nanodisks [30] or quantum dots [31]. On the other hand, PDs based on the increase of thin film optical absorption by the use of an optical microcavity have been already reported in literature with the name of resonant cavity enhanced (RCE) photodetectors [32]. RCE PDs are able to shrink the optical field inside the cavity within the active intrinsic layer of III-V PIN diodes [32] allowing reducing the size of the absorbing layer, and consequently the carrier transit time, without degradation of the device efficiency.

Taking advantage of this idea, in this work we propose a new concept of Si-based RCE PDs operating at 1550 nm where graphene/Si Schottky junctions have been incorporated into a Fabry–Perot microcavity [33] that could be realized starting from a crystalline-Si (c-Si) based distributed Bragg reflector (DBR) substrate. Indeed, thanks to a double silicon on insulator (SOI) process, DBRs consisting of two periods of c-Si/SiO$_2$ can be realized and optimized for high reflectivity around 1550 nm [34]; they are named double-SOI (DSOI). As second mirror two options have been considered: a distributed Bragg reflector constituted by some periods of silicon nitride (Si$_3$N$_4$)/hydrogenated amorphous silicon (a-Si:H) and a metallic reflector based on a thick layer of gold. In more detail, our proposal is to replace the III-V P-I-N diodes used in classical RCE configuration by a buffer layer added to a graphene/silicon Schottky junction. The buffer layer is useful to accommodate the localized optical field on the thin graphene layers where charge carriers are excited by photons and then emitted into c-Si through the Schottky junction making detection possible in the NIR spectrum. As buffer layer, we have chosen a-Si:H, a material which can be deposited at low temperature by a plasma-enhanced chemical vapor deposition (PECVD) system. Moreover, a-Si:H is characterized by a refractive index very close to that of c-Si at 1550 nm which is mandatory to reduce the Fresnel reflection at the interface and, consequently, to consider the a-Si:H/graphene/c-Si three-layer structure as one unique optical cavity. Finally, we have investigated the possibility to collect the current both longitudinally and transversally to the direction of the incoming light putting in evidence advantages and disadvantages of each configuration.

2. Photodetector Performance and Theoretical Background

It is well-known that a very important figure of merit for a PD is the internal quantum efficiency (IQE) η_{int}, defined as the number of charge carriers collected per absorbed photon. IQE is linked to the external quantum efficiency (EQE) η_{ext} (number of charge carriers collected per incident photon) by the following formula: $\eta_{ext} = A \, \eta_{int}$, where A is the optical absorption of the active material. A macroscopic measurable magnitude is the responsivity R, i.e., the ratio of the photogenerated current (I_{ph}) to the incident optical power (P_{inc}). The responsivity R is linked to EQE η_{ext} by the following formula:

$$R = \frac{I_{ph}}{P_{inc}} = \frac{\lambda(nm)}{1242}\eta_{ext} = \frac{\lambda(nm)}{1242}A\,\eta_{int} \tag{1}$$

As reported in literature, the internal quantum efficiency of an IPE-based graphene/silicon Schottky photodetectors is given by the following [35]:

$$\eta_{int} = \frac{1}{2}\frac{(h\nu)^2 - (q\phi_B)^2}{(h\nu)^2} \tag{2}$$

where $h\nu$ is the photon energy, $q = 1.602 \times 10^{-19}$ C is the charge electron and $q\phi_B$ is the Schottky barrier height of the graphene/silicon junction. It is well-known that due to the Fermi level shift in graphene, the Schottky barrier of a graphene/silicon junction lowers by increasing the reverse voltage applied. Of course, the decrease in ϕ_B leads to increased responsivity, thus the effects of the reverse voltage V_R on the responsivity can be investigated. In other words, the Schottky barrier $q\phi_B$ can be viewed as the sum of the Schottky barrier at zero-bias $q\phi_{B0}$ and the Schottky barrier lowering $q\Delta\phi_B(V_R)$ due to the increase in reverse voltage: $q\phi_B(V_R) = q\phi_{B0} + q\Delta\phi_B(V_R)$. To this aim we take advantage of the work of Tongay et al. [36] where, the Schottky barrier lowering $\Delta\phi_B(V_R)$ due to the Fermi level shift has been calculated by the following formula:

$$q\Delta\phi_B(V_R) = -\frac{1}{2}h\nu_F\sqrt{\frac{\pi\varepsilon_s\varepsilon_0 N(V_{bi} + V_R)}{2qn_0}} \tag{3}$$

where n_0 is the graphene extrinsic doping (a typical value is 5×10^{12} cm^{-2} [36]), N is the semiconductor doping, $v_F = 1.1 \times 10^8$ cm/s is the Fermi velocity, $\hbar = 6.5 \times 10^{-16}$ eVs is the Plank constant, $\varepsilon_0 = 8.859 \times 10^{-14}$ C/cmV is the permittivity of vacuum, $\varepsilon_S = 11.4$ is the relative permittivity of Si and V_{bi} is the built-in potential of the junction.

Device bandwidth is another figure of merit very important for a PD, in particular in telecom and datacom applications. The main factors limiting the time response of a PD integrated into an optical microcavity are [37,38]: (1) the carriers transit time τ_{tr} across the charge spatial region, (2) the charge/discharge time τ_{RC} linked to both the junction capacitance C_j and the load resistance R_L; (3) the cavity photon lifetime τ_{ph} [39]. Thus, the overall time constant of the PD is $\tau = \tau_{tr} + \tau_{RC} + \tau_{ph}$. For the $\tau_{RC} = R_L C_j$ calculation, we can consider $R_L = 50$ Ω (typical value for high-speed applications) and the junction capacitance given by the following formula: $C_j = A_{PD} \cdot \varepsilon_0 \cdot \varepsilon_S / W$, where $A_{PD} = \pi r^2$ is the circular graphene area with radius r in contact with Si and W is the charge spatial region width On the other hand, the cavity photon lifetime can be calculated by the following formula $\tau_{ph} = 1/2\pi\delta v$ [39], being δv the spectral width of the absorption peak at half maximum. The transit time τ_{tr} can be written as $\tau_{tr} = t/v_{sat}$ where t is the maximum distance that the electron must travel before being collected (by considering this distance completely depleted) and v_{sat} is the carrier saturation velocity in Si. As we will see, this distance t strongly affects the bandwidth of the device.

Finally, the cut-off frequency can be estimated as:

$$f_{3dB} = \frac{1}{2\pi\tau} = \frac{1}{2\pi\left(\frac{t}{v_{sat}} + \frac{\pi\varepsilon_0\varepsilon_S R_L}{W}r^2 + \frac{1}{2\pi\delta v}\right)} \tag{4}$$

Another very important figure of merit is the noise equivalent power (NEP), i.e., the minimum optical power which can be detected by a PD, in an approximated form, which can be written as:

$$NEP = \frac{\sqrt{2qJ_d}}{R} \tag{5}$$

Being q the electron charge and J_d the dark current density that, for Schottky PD, can be written as:

$$J_d = A^* T^2 e^{-\frac{q\phi_{B0}}{kT}} \tag{6}$$

where A^* is the Richardson constant (32 A/cm^{2}K^2 for p-Si [37]), T the absolute temperature, k the Boltzmann constant and $q\Phi_{B0}$ is the Schottky barrier of the graphene/silicon Schottky junction at zero bias. The unity of measure of NEP is $W/cm\sqrt{Hz}$.

3. Photodetector Concept: From the Idea to the Device

In this section the basic idea of the device will be presented, some details on the numerical simulations will be provided, and the materials selected for a possible fabrication will be discussed.

3.1. Si-Based Resonant Cavity Enhanced (RCE) Photodetectors

Classical RCE PDs are able to concentrate the enhanced optical field in the absorbing intrinsic region of a PIN diode realized by a III-V semiconductor [32], where the P, I and N are characterized by a slightly different stoichiometry in such a way as to neglect the reflections at the interface and to consider the whole PIN structure as an unique cavity.

In our proposed device, the P-I-N structure has been replaced by a Schottky graphene/c-Si Schottky junction on which is added a buffer layer useful to accommodate the localized optical field on graphene as shown in Figure 2. As buffer layer we have chosen hydrogenated amorphous silicon (a-Si:H) because it can both be deposited at low temperature by a PECVD system and its refractive index at 1550 nm is 3.58 [40], very close to that one of c-Si (3.48) [41]. This latter property together with the high transparence of the graphene layer allow considering the whole a-Si:H/graphene/c-Si three-layer structure as one unique optical cavity with negligible reflections at the interfaces.

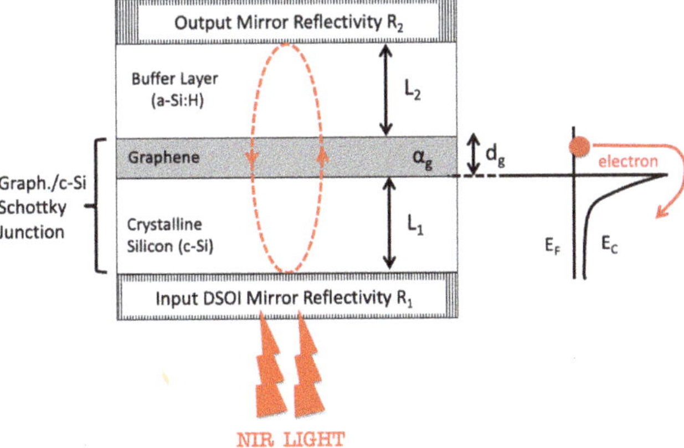

Figure 2. Simplified model of the proposed Fabry–Pérot graphene/silicon Schottky photodetector.

It is widely reported in literature that the maximum absorption of RCE PDs occurs when the reflectivity of the output mirror is close to the unity. This is the reason why we consider illumination from the bottom as shown in Figure 2, indeed as will see in Section 3.3 the DSOI DBR is characterized by a limited reflectivity making it not suitable to work as output mirror. It is worth mentioning that an optimized RCE structure is characterized by an output mirror reflectivity $R_2 = 1$ and an input mirror reflectivity $R_1 = R_2 \times e^{-\alpha_g d_g}$, being α_g and d_g the absorption coefficient and thickness of the graphene absorbing layer, respectively [35]. All numerical simulations have been carried out by the transfer matrix method (TMM) taking into account the dispersion curve shown in Figure 3a–d.

Figure 3b shows that only graphene and Au are provided of a non-negligible extinction coefficient (absorption coefficient) in the range of wavelength taken into account. Moreover, Figure 3d shows that non-negligible absorption appears when Si is considered heavily doped due to free carrier absorption. In the next, for heavily and lightly doped silicon semiconductor we'll intend doping of 5×10^{19} cm^{-3} and 1×10^{15} cm^{-3}, respectively.

All dispersion curves shown in Figure 3 have been taken by references [41–43], while the graphene complex refractive index n_g can be obtained by [44]:

$$n_g = \sqrt{\varepsilon_g} = \sqrt{2.148 + j\frac{G\lambda}{2d_g}} \tag{7}$$

where ε_g is the relative permittivity of graphene, λ is the wavelength, $d_g = 0.335$ nm is the graphene thickness and $G = \frac{q^2}{2\varepsilon_0 hc} = 0.0073$ is the fine structure constant [45] (being h = 6.626×10^{-34} Js the Planck constant, and c = 3×10^{10} cm/s the speed of light in vacuum).

Finally, the heavily-doped c-Si complex refractive index has been calculated by applying the theory of the free carrier absorption [46]. The calculation provides the variation of both the real part of the refractive index Δn_{Si} and the absorption coefficient $\Delta \alpha_{Si}$ of doped c-Si as a function of the donor and acceptor concentration atoms, N_d and N_a, respectively [46]:

$$\Delta n_{Si} = \frac{q^2 \lambda^2}{8\pi^2 c^2 \varepsilon_0 n^0_{Si}}\left(\frac{N_d}{m^*_e} + \frac{N_a}{m^*_h}\right) \tag{8}$$

$$\Delta\alpha_{Si} = \frac{q^3\lambda^2}{4\pi^2 c^3 \varepsilon_0 n_{Si}^0}\left(\frac{N_d}{\mu_e(m_e^*)^2} + \frac{N_a}{\mu_h\left(m_h^*\right)^2}\right) \tag{9}$$

where n_{Si}^0 is the refractive index of unperturbed crystalline Si, m_e^* and m_h^* are the conductivity effective mass of electrons and holes, respectively, while μ_e and μ_h are the electrons and holes mobility, respectively. It is worth remembering that the extinction coefficient κ_{Si} can be derived by the absorption coefficient α_{Si} by the following formula: $\alpha_{Si} = (4\pi/\lambda)\,\kappa_{Si}$.

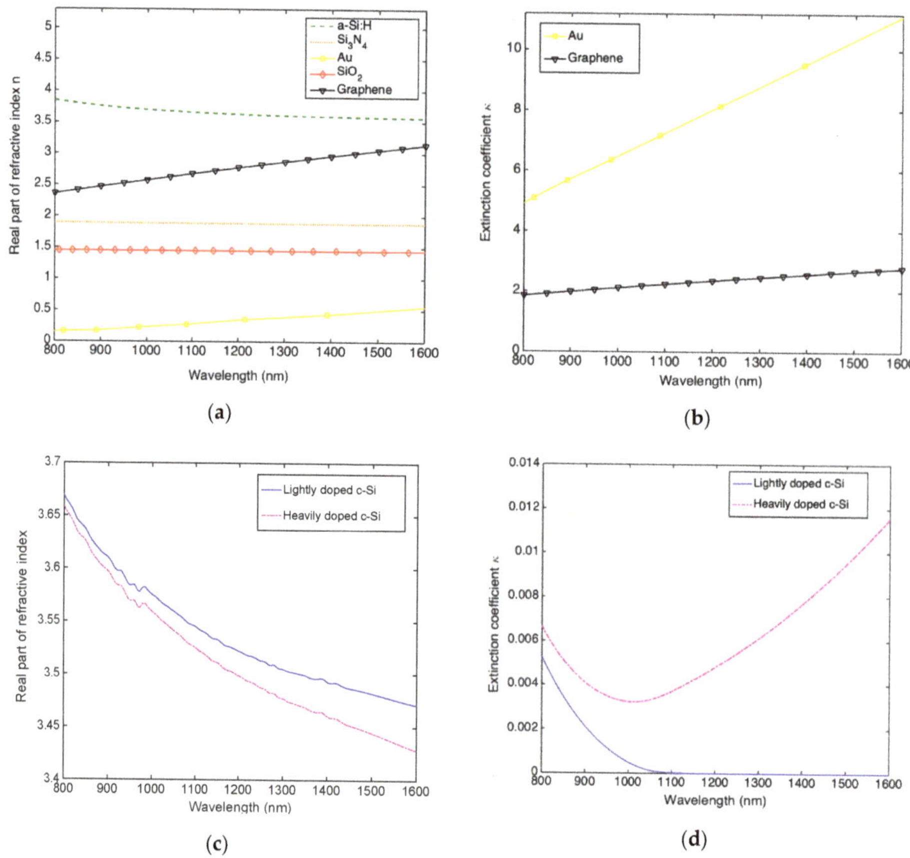

Figure 3. Dispersion curves of all materials used in our numerical simulation: (**a**) real part of refractive index and (**b**) extinction coefficient of a-Si:H, Si$_3$N$_4$, Au, SiO$_2$ and graphene. (**c**) real part of refractive index and (**d**) extinction coefficient of lightly (1×10^{15} cm^{-3}) and heavily doped (5×10^{19} cm^{-3}) c-Si.

3.2. Buffer Layer

By moving our attention on the cavity, as already mentioned, we need to choose a material working as buffer layer which can deposited on graphene. The a-Si:H has been already successfully deposited on graphene without damaging it [47], in addition this material is characterized by a refractive index very close to that of c-Si at 1550 nm. This property combined with the high transparency of graphene lead to negligible Fresnel reflection at the a-Si:H/graphene and c-Si/graphene interfaces. Indeed, in literature has been already proved that the a-Si:H/graphene/c-Si three-layer cavity can be modelled by the classical theory of RCE PDs [40].

Figure 4 shows the reflections at the c-Si/graphene and a-Si:H/graphene interfaces. Low reflections in the order of 10^{-4} are reported at 1550 nm.

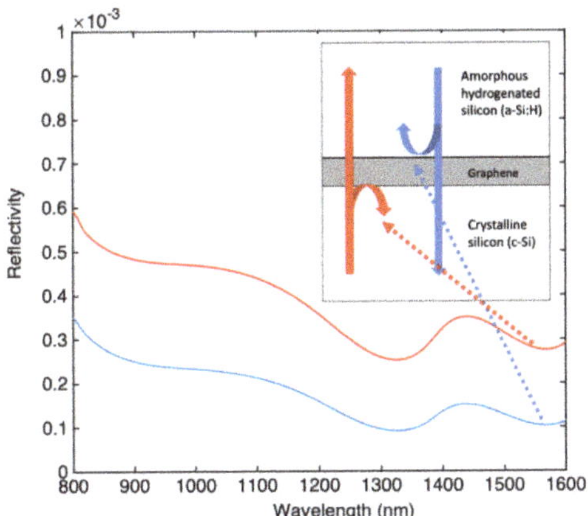

Figure 4. Reflectivity inside the cavity at the c-Si/graphene (red) and a-Si:H/graphene (blue) interfaces.

3.3. Output and Input Mirror of the Fabry–Pérot Microcavity

As output mirror of the Fabry-Pérot microcavity we take into account two options: a DBR constituted by alternating layer of a-Si:H/Si$_3$N$_4$ (that can be deposited at low temperature by a PECVD system) and a metal reflector (MR) constituted by a thick Au metallic layer. It is well-known that a DBR consists of a multilayer-stack of alternate high- and low-refractive index layers, all one quarter wavelength thick. In order to get high reflectivity at 1550 nm by a Si$_3$N$_4$/a-Si:H DBR the thicknesses are calculated as high as 213 and 108 nm, respectively. Figure 5a shows the DBR reflectivity for 3, 4 and 5 pairs of Si$_3$N$_4$/a-Si:H.

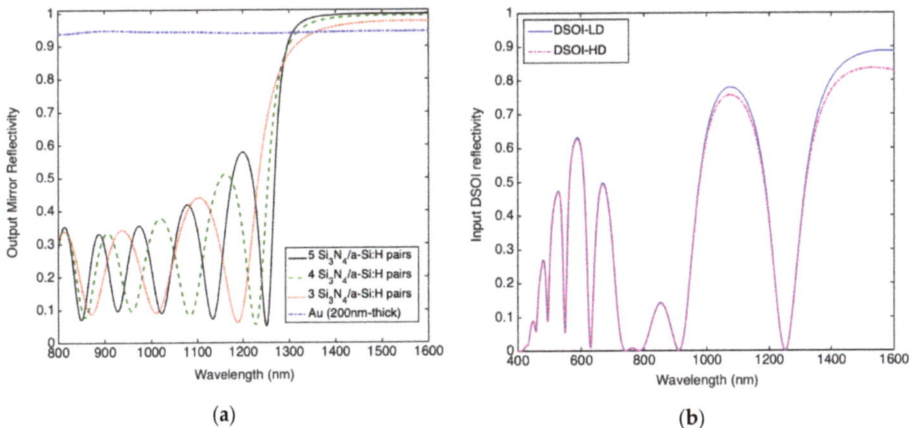

Figure 5. Reflectivity vs wavelength of the: (**a**) output mirror realized by 200 nm-thick Au metal reflector (MR) and distributed Bragg reflector (DBR) constituted by 3, 4 and 5 Si$_3$N$_4$/a-Si:H pairs and (**b**) input double silicon on insulator (DSOI) mirror constituted by the first c-Si layer both lightly and heavily doped.

In addition, Figure 5a shows the reflectivity of a 200 nm-thick Au layer. It is well-known that metals are characterized by a plasma frequency higher than frequency in the optical spectrum, this inhibits the optical propagation in the metal leading to high reflectivity.

Moving our attention to the input mirror, the possibility to fabricate a DBR by alternating layers of silicon dioxide (SiO_2) and c-Si is reported in reference [34]. DBR constituted by alternating layers of c-Si/SiO_2 are characterized by a large refractive index contrast (3.48/1.47 at 1550 nm, respectively) allowing the realization of high-reflectivity, wide spectral stop-band DBR made of few periods [48]. These DBRs are realized by a double silicon on insulator process, thus they are constituted by two c-Si/SiO_2 pairs and named DSOI. Because the manufacturing process typically does not allow the manufacture of Si thickness as thin as $\lambda/4n_{Si}$, a c-Si thickness of $3\lambda/4n$ is typically used [48]. A further advantage of this structure is that on top of the reflector there is a crystalline layer of Si which can be used for growing (by epitaxial processes) other crystalline Si layers with different doping, for instance for realizing heavily doped layers necessary for the fabrication of Ohmic contacts. We have investigated c-Si/SiO_2 DSOI with thicknesses of 340 nm/270 nm in two configurations: one with the first c-Si layer heavily doped (HD) and the other with the first c-Si layer lightly doped (LD). We name them DSOI-HD and DSOI-LD, respectively. For the DSOI reflectivity calculation, c-Si has been considered as both input and output semi-infinite medium. In Table 1, reflectivity and thicknesses of all reflectors discussed in this section are reported.

Table 1. Reflectivity and thicknesses of all investigated reflectors.

Reflector	Mirror	Reflectivity at 1550 nm	Thickness
DSOI-LD	Input	0.8790	340 nm (Si) and 270 nm (SiO_2)
DSOI-HD	Input	0.8344	340 nm (Si) and 270 nm (SiO_2)
DBR (3 pairs of Si_3N_4/a-Si:H)	Output	0.9756	213 nm (Si_3N_4) and 108 nm (a-Si:H)
DBR (4 pairs of Si_3N_4/a-Si:H)	Output	0.9932	213 nm (Si_3N_4) and 108 nm (a-Si:H)
DBR (5 pairs of Si_3N_4/a-Si:H)	Output	0.9985	213 nm (Si_3N_4) and 108 nm (a-Si:H)
Au	Output	0.9451	200 nm

4. Results

This section will show the results of the numerical simulations carried out by TMM [49] implemented by custom codes written in Matlab. Devices can be realized in two configurations as shown in Figure 6a,b.

In particular, Figure 6a shows a structure where the first layer of the DSOI is lightly doped while only a small region placed under the collecting metal is heavily doped for getting an Ohmic contact (DSOI-LD). In other words, in this configuration we can say that both Ohmic and Schottky contacts are realized on the same plane and photoexcited charge carriers emitted by graphene into c-Si are collected transversally to the direction of the incoming light. We name this configuration: the transverse collection device. As should be noted, in this configuration the maximum distance t that a charge carrier generated in the center of the graphene disk has to cover before being collected is further high because the radius of the graphene area is in the order of some tens of μm. As consequence, even if all this distance t is completely depleted the slow carrier transit time τ_{tr} is expected to reduce the device bandwidth.

On the other hand, in Figure 6b is shown a structure where the first layer of the DSOI is entirely heavily doped (DSOI-HD). In other words, in this configuration the Ohmic contact is placed in front of the Schottky contact and the photoexcited charge carriers emitted by graphene into c-Si are collected in parallel (longitudinally) to the direction of the incoming light. We name this configuration: longitudinal collection device. It should be noted in this configuration that the distance t that any charge carrier emitted by graphene into c-Si has to cover before being collected is the thickness of the c-Si layer composing the cavity. This value is in the order of some hundreds of nm, as a consequence the carrier transit time is two orders of magnitude lower with respect to transverse collection devices shown in Figure 6a. However, the heavily doped layer in the DSOI reflector absorbs part of the light trapped in the cavity at any round-trip, thus in this configuration a reduced graphene absorption, and consequently responsivity, is expected.

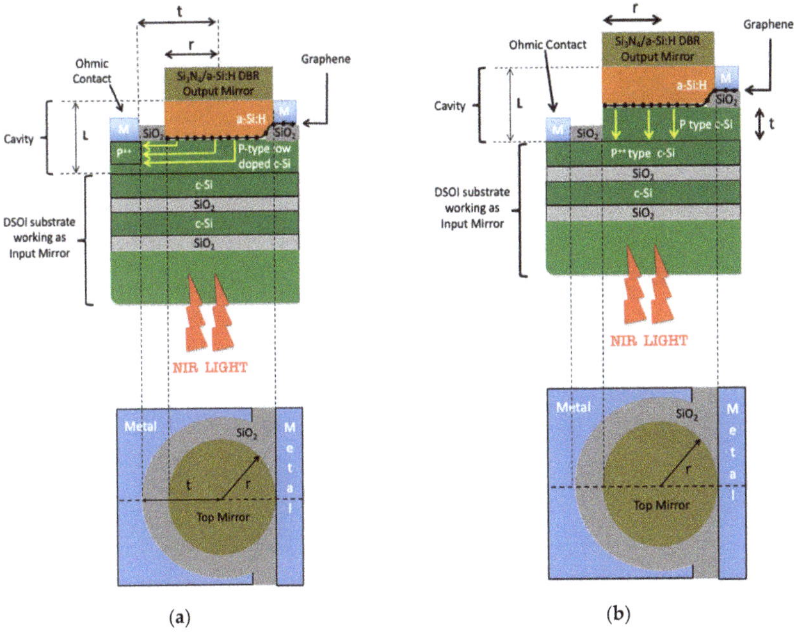

Figure 6. Sketch of the Fabry–Pérot graphene/Si Schottky PD in the (**a**) transverse and (**b**) longitudinal collection configuration.

4.1. Transverse Collection Configuration (TCC)

In this section we investigate the transverse collection configuration (TCC) shown in Figure 6a. As output mirror, a DBR constituted of 3, 4 and 5 Si_3N_4/a-Si:H pairs is considered. On the other hand, the input mirror of the device is constituted by a DSOI-LD. Of course, as shown in Figure 6a, the optical microcavity is formed by a-Si:H/graphene/c-Si three-layer structure.

We have performed numerical simulations in order to calculate the graphene absorption at 1550 nm by varying the thicknesses of both a-Si:H and c-Si layers comprising the cavity for a DBR output mirror composed of 3, 4 and 5 Si_3N_4/a-Si:H pairs. Results are shown in Figure 7a–c, respectively; because the position of the maximum of the standing wave inside the cavity does not depend on the reflectivity of two mirrors, in any case that the maximum graphene absorption can be obtained for 111 nm-thick and 214 nm-thick of c-Si and a-Si:H, respectively.

The spectral graphene absorption around 1550 nm for the optimized thicknesses is shown in Figure 7d. Figure 7d shows that the maximum graphene absorption is 0.44, 0.54 and 0.58 while the full width at half maximum (FWHM) are 10.17 nm, 9 nm and 8.54 nm, for DBRs composed by 3, 4 and 5 Si_3N_4/a-Si:H pairs, respectively. Of course the maximum absorption is obtained for the cavity characterized by the highest finesse, i.e., that one provided of a DBR constituted by 5 pairs of Si_3N_4/a-Si:H.

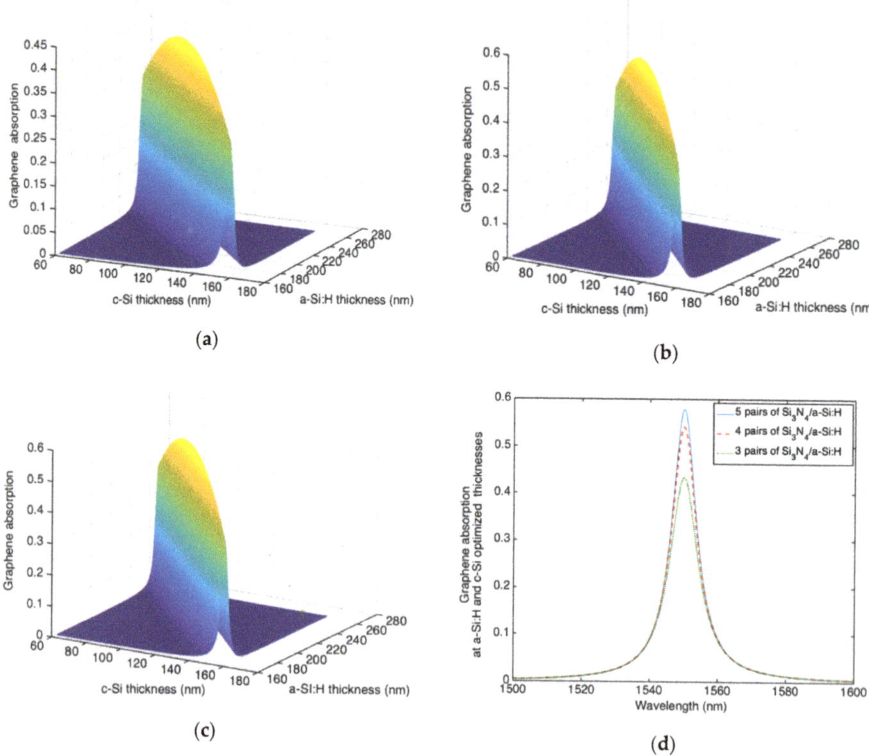

Figure 7. Graphene absorption as function of both c-Si and a-Si:H thicknesses for the transverse collection configuration (TCC) Fabry–Pérot graphene/Si Schottky PD provided of a DBR output mirror constituted by (**a**) 3, (**b**) 4 and (**c**) 5 pairs of Si$_3$N$_4$/a-Si:H and (**d**) spectral graphene absorption at the optimized c-Si and a-Si:H thicknesses.

Taking advantage of the calculated graphene absorption, by applying Equations (1) and (2), we have calculated the spectral responsivity at zero bias.

As shown in Figure 8a the maximum responsivity at 1550 nm is 0.19 A/W, 0.23 A/W and 0.24 A/W for DBR composed of 3, 4 and 5 Si$_3$N$_4$/a-Si:H pairs, respectively.

In order to verify if a further increase in responsivity at 1550 nm can be obtained by increasing the reverse bias, we use Equations (1)–(3). Figure 8b shows a very limited increase in responsivity also at −10 V of reverse bias applied, leading to the idea that these devices could also work at low reverse voltage without degrading their efficiency.

Moving our attention on the bandwidth of the device, Figure 9a–c show the time constants discussed in the Section 2 as function of the radius r of the graphene active area, for DBR composed by 3, 4 and 5 Si$_3$N$_4$/a-Si:H pairs, respectively. In addition, Figure 9a–c show the 3 dB roll-off frequency as function of the graphene disk radius r, too.

Figure 9a–c have been calculated by considering: (i) for the τ_{tr} calculation, a $v_{sat} = 10^7$ cm/s [37] and a drift length $t = r$; (ii) for the $\tau_{RC} = R_L C_j$ calculation a load resistance $R_L = 50 \ \Omega$ and a junction capacity $C_j = (\pi r^2 \varepsilon_0 \varepsilon_s)/W$, being $W = \sqrt{((2 \cdot \varepsilon_0 \cdot \varepsilon_s)/qN_a)} \cdot V_{bi} = 0.5 \ \mu m$ the length of the depletion layer that has been evaluated by considering a built-in potential $V_{bi} = \Phi_{B0} - (E_F - E_V) = 0.196$ V (with the Schottky barrier $\Phi_{B0} = 0.45$ V [50] and the difference between the extrinsic Fermi level and the Si valence band $E_F - E_V = 0.254$ V calculated starting from a p-type doping $N_a = 10^{15}$ cm^{-3}; (iii) the cavity photon lifetime $\tau_{ph} = 1/2\pi\delta\nu$, being $\delta\nu$ the spectral width of the absorption peak which can be obtained by the FWHM extracted by Figure 7d and converted into frequencies leading to: δv = 1270, 1124 and 1066 GHz for DBR composed by 3, 4 and 5 Si$_3$N$_4$/a-Si:H pairs, respectively.

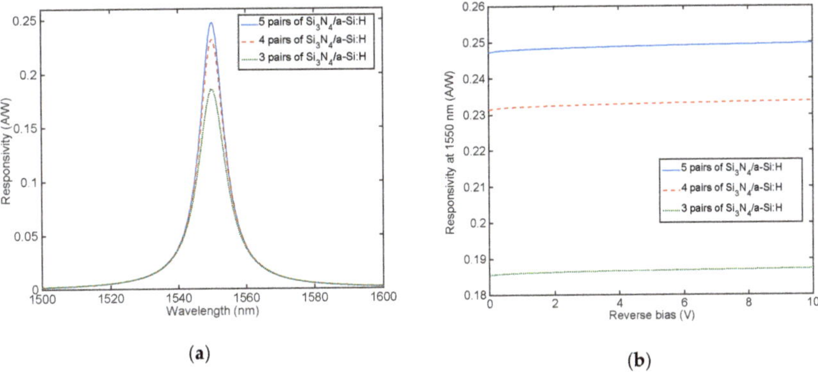

(**a**)

(**b**)

Figure 8. Responsivity (**a**) as function of the wavelength and (**b**) at 1550 nm as function of the reverse voltage applied for the TCC Fabry–Pérot graphene/Si Schottky PD.

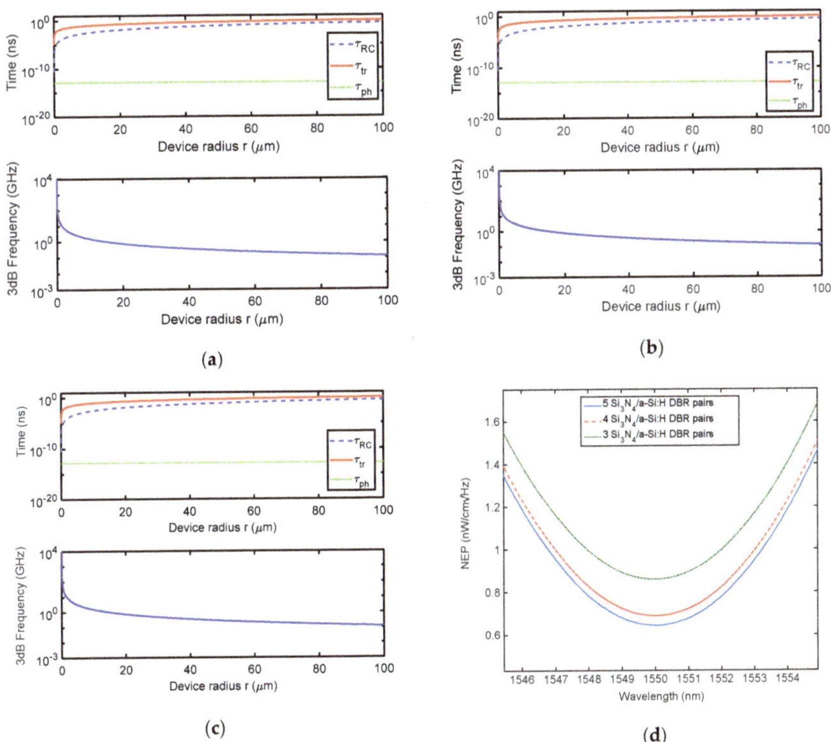

(**a**)

(**b**)

(**c**)

(**d**)

Figure 9. Carriers transit time τ_{tr}, charge/discharge time τ_{RC}, cavity photon lifetime τ_{ph} and 3 dB roll-off frequency versus graphene disk radius r for the TCC Fabry-Pérot graphene/Si Schottky PD provided of a DBR output mirror constituted by (**a**) 3, (**b**) 4 and (**c**) 5 pairs of Si$_3$N$_4$/a-Si:H and (**d**) spectral noise equivalent power (NEP).

Figure 9a–c show that in this configuration the limiting factor is the transit time; of course, by increasing the radius r, the τ_{RC} constant time grows in a square way approximating the value of the transit time (which instead depends on a linear way from the radius r). Figure 9a–c show that the transverse collection configuration is able to work above 1 GHz if the radius r of the graphene active area is lower than 18 µm making harder the optical coupling with the incoming radiation.

Finally, Figure 9d shows the device NEP for DBRs composed by 3, 4 and 5 Si_3N_4/a-Si:H pairs. NEP has been calculated by Equations (5) and (6) (with $A^* = 32$ A/cm^{2}K^2, $T = 300$ K, $k = 8.617 \times 10^{-5}$ eV/K and $q\Phi_{B0} = 0.45$ eV) and by taking into account the results shown in Figure 8a. NEP decreases by increasing the finesse of the cavity due to the increase in responsivity, the minimum NEP at 1550 nm is 0.6 W/cm $\sqrt{Hz}$ for a DBR with 5 Si_3N_4/a-Si:H pairs.

4.2. Longitudinal Collection Configuration (LCC)

In this section we investigate the longitudinal collection configuration (LCC) shown in Figure 6b. As output mirror, a DBR constituted of 3, 4 and 5 Si_3N_4/a-Si:H pairs is considered.

On the other hand, the input mirror of the device is constituted by a DSOI-HD. Of course, as also shown in Figure 6a, the optical microcavity is formed by a-Si:H/graphene/c-Si three-layer structure.

We have performed numerical simulations in order to calculate the graphene absorption at 1550 nm by varying the thicknesses of both a-Si:H and c-Si layers comprising the cavity for DBRs with 3, 4 and 5 Si_3N_4/a-Si:H pairs. The maximum graphene absorption can be achieved when the thickness of c-Si and a-Si:H are 114.9 nm and 214.0 nm, respectively. The spectral graphene absorption around 1550 nm for these optimized thicknesses is shown in Figure 10a for DBRs constituted of 3, 4 and 5 Si_3N_4/a-Si:H pairs.

Figure 10a shows that the maximum graphene absorption is 0.24, 0.28 and 0.29 while the FWHM are 13.42 nm, 12.43 nm and 12.13 nm, for DBRs with 3, 4 and 5 Si_3N_4/a-Si:H pairs, respectively. By contrast the DSOI mirror, due to its first heavily doped c-Si layer, is characterized by a free carrier absorption of 0.53, 0.62 and 0.66 for DBRs with 3, 4 and 5 Si_3N_4/a-Si:H pairs, respectively. For this reason, the graphene optical absorption is lower than that one reported for transverse collection configuration.

Even if Figure 3b,d show that the absorption coefficient of graphene is higher than heavily doped c-Si, the latter is much thicker (340 nm-thick) with respect to graphene (0.335 nm-thick), thus absorbing the most part of the light trapped into the cavity. By applying Equations (1) and (2) we can achieve the spectral responsivity at zero bias. As shown in Figure 10c the maximum responsivity at 1550 nm is 0.10 A/W, 0.12 A/W and 0.13 A/W for DBRs with 3, 4 and 5 Si_3N_4/a-Si:H pairs, respectively. In order to verify if a further increase in responsivity at 1550 nm can be obtained by increasing the reverse bias we use Equations (1)–(3), also in this case the increase in responsivity at −10 V is very limited, as reported in Figure 10d.

Moving our attention to the bandwidth of the device, the time constants discussed in Section 2 have been calculated as already described for TCC. The only difference concerns the calculation of both the transit time $\tau_{tr} = t/vsat$, being t the thickness of the c-Si layer composing the cavity ($t = 114.9$ nm) and the cavity photon lifetime $\tau_{ph} = 1/2\pi\delta v$ which is expected to reduce due to the increase in δv associated to the increased cavity losses. The frequency spectral widths δv have been calculated by Figure 10a as 1675, 1552 and 1515 GHz for DBR with 3, 4 and 5 Si_3N_4/a-Si:H pairs, respectively. Figure 11a shows three time constants and 3 dB roll-off frequency for a device provided of a DBR constituted by 5 Si_3N_4/a-Si:H pairs. Due to the reduced transit time in LCC, Figure 11a shows as the limiting factor is now the RC time constant. Because the junction capacity C is linked to the graphene area in contact with Si, by reducing the area an increase in bandwidth is expected, on the other hand, a smaller area could make harder the optical coupling of the incoming radiation. It is worth noting that in this configuration not only the cavity photon lifetime but also the transit time are independent of the radius r of the graphene active area. Figure 11a shows that if the radius of the active area is 70 µm the LCC is able to work at 1 GHz while at the same radius the TCC is characterized by a bandwidth of only 186 MHz (see Figure 9c).

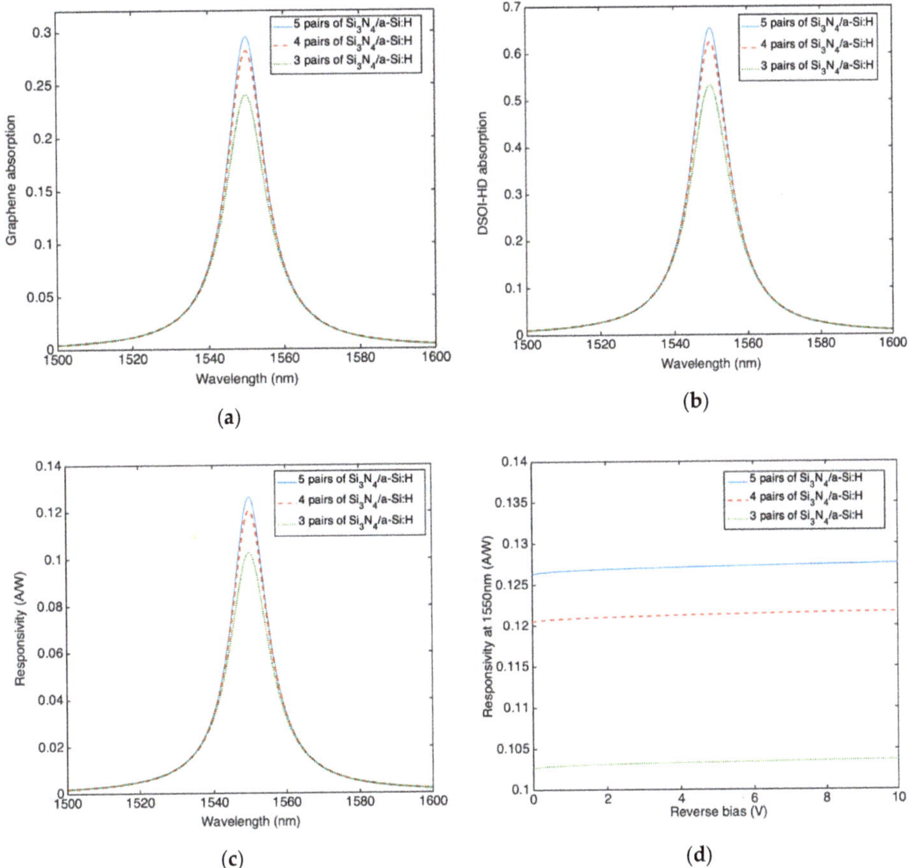

Figure 10. (**a**) Graphene optical absorption, (**b**) DSOI optical absorption, (**c**) responsivity as function of the wavelength for the longitudinal collection configuration (LCC) Fabry–Pérot graphene/Si Schottky PD provided of a DBR output mirror constituted by 3, 4 and 5 pairs of Si$_3$N$_4$/a-Si:H and (**d**) responsivity at 1550 nm as function of the reverse bias.

Finally, Figure 11b shows the device NEP for DBRs composed of 3, 4 and 5 Si$_3$N$_4$/a-Si:H pairs. NEP has been calculated by Equations (5) and (6) as already discussed for the TCC. NEP decreases by increasing the finesse of the cavity due to the increase in responsivity, the minimum NEP at 1550 nm is 1.26 W/cm $\sqrt{\text{Hz}}$ for a DBR with 5 pairs of Si$_3$N$_4$/a-Si:H.

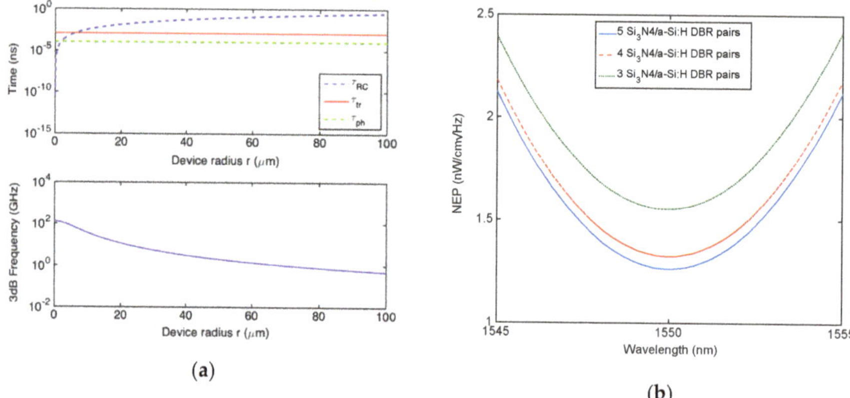

Figure 11. (**a**) Carriers transit time τ_{tr}, charge/discharge time τ_{RC}, cavity photon lifetime τ_{ph} and 3 dB roll-off frequency versus graphene disk radius r for the LCC Fabry–Pérot graphene/Si Schottky PD provided of a DBR output mirror constituted by 5 pairs of Si$_3$N$_4$/a-Si:H and (**b**) spectral NEP.

5. Discussion

In this section we put in comparison the optimized transverse and longitudinal collection Fabry–Pérot graphene/Si Schottky PD. As the output mirror is not only considered a DBR constituted by Si$_3$N$_4$/a-Si:H (5 pairs for optimized structures) but also a 200 nm-thick gold Au MR which could be a good option for reducing the manufacturing complexity. However, the metallic mirror absorbs part of the light trapped in the cavity at any round-trip reducing the graphene absorption, consequently a reduced responsivity is expected with respect to the counterpart based on DBR.

Figure 12a shows a comparison of the spectral graphene absorption for the four structures (transverse and longitudinal collection configuration with both DBR and MR as output mirror); as expected, the maximum graphene absorption is obtained for the configuration which does not involve other absorbing layers apart from graphene. As a consequence, Figure 12b shows that the TCC is characterized by the highest responsivity of 0.24 A/W which is a very interesting value mainly by considering that these Si-based PDs could be monolithically integrated with an electronic circuitry and not separately fabricated and then assembled as happened for the fabrication of NIR imaging systems based on InGaAs or germanium. As shown in Figure 12d, the same configuration is also characterized by the lowest NEP of 0.6 nW/cm $\sqrt{\text{Hz}}$ and this is why this configuration could be preferred for applications where high sensitivities are required, for instance in free space optical communications in both spatial and terrestrial environment.

By contrast, Figure 12c shows as the longitudinal configuration is characterized by higher bandwidth than transverse counterpart. This is due to the reduced transit time which make the RC time constant the limiting factor. In LCC, a further increase in bandwidth could be obtained by reducing the RC time constant, for instance by reducing the graphene area in contact with silicon, i.e., by reducing the radius r. However, it is worth mentioning that if the active area becomes too small more complex optical coupling techniques are required for focusing the radiation on the active area. For a radius r = 70 µm, longitudinal structures provided by both DBR and MR output mirrors, are characterized by a bandwidth of 1 GHz while the transverse one by a bandwidth of only 186 MHz. LCC could be used for applications where the high speed is the main requirement.

Figure 12a,b,d show that from a point of view of graphene absorption, responsivity and NEP, the LCC provided of DBR as output mirror is almost equivalent to the TCC provided of MR as output mirror, thus for applications where high bandwidth is not the main requirement, the transverse structure could be preferred because characterized by a lower manufacturing complexity.

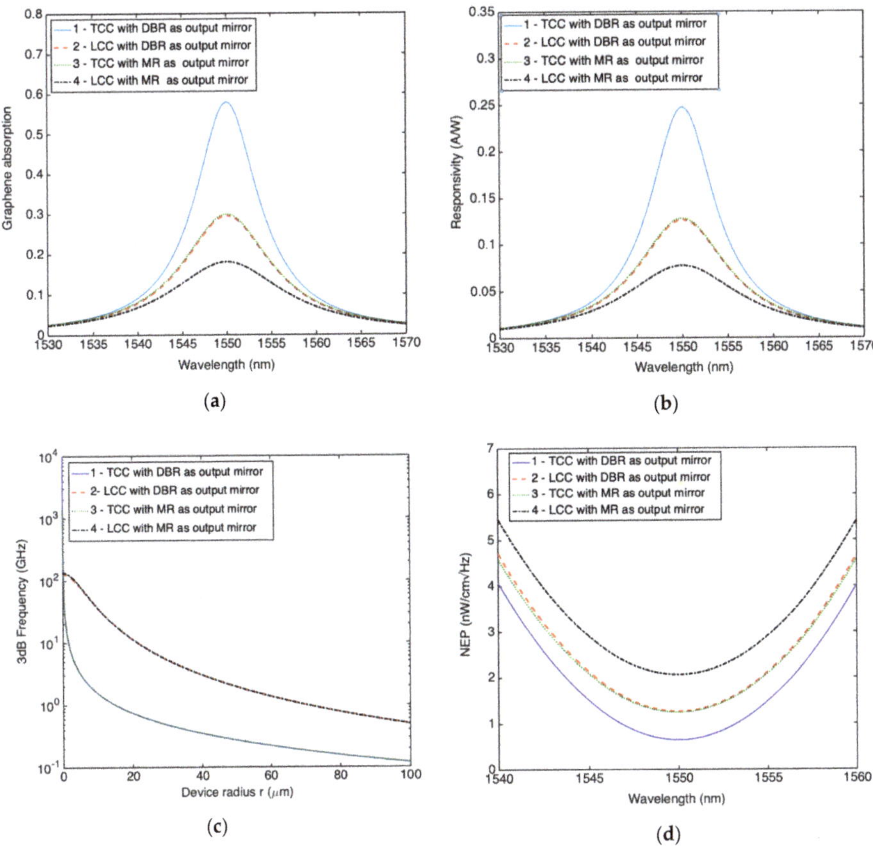

Figure 12. Comparison between optimized TCC and LCC Fabry-Pérot graphene/Si Schottky PDs provided of both DBR and MR as output mirror in term of: (**a**) spectral graphene absorption, (**b**) spectral responsivity, (**c**) 3 dB frequency and (**d**) spectral NEP.

The performance of any configuration and related optimization parameters are reported in Table 2. Table 2 shows that concerning the structures provided of DBR as output mirror, the longitudinal configuration is characterized by the highest responsivity × bandwidth parameter, while the transversal one is characterized by the narrowest FWHM, i.e., by the highest selectivity. By contrast, the lowest selectivity is shown by the longitudinal configuration provided of MR as output mirror due to the highest losses in the cavity given by both the metal reflector and the doped DSOI.

Table 2. Summary of the main performance of the optimized TCC and LCC Fabry–Pérot graphene/Si Schottky PD provided by both DBR and MR as output mirror.

Input Mirror	Output Mirror	c-Si Thick nm	a-Si:H Thick nm	FWHM nm	Resp. at 1550 nm A/W	3 dB Freq. at $r = 70\ \mu m$ MHz	Resp. × 3 dB Freq. A/WxMHz	NEP at 1550 nm $nW/cm\sqrt{Hz}$
1-TCC DSOI-LD	DBR (5 pairs)	111.0	214.0	8.54	0.24	186	44.6	0.60
2-LCC DSOI-HD	DBR (5 pairs)	114.9	214.0	12.13	0.13	1000	126.0	1.26
3-TCC DSOI-LD	Metal	110.1	303.7	12.18	0.13	186	23.8	1.24
4-LCC DSOI-HD	Metal	110.1	307.4	15.59	0.0775	1000	77.5	2.10

6. Conclusions

In this work we have theoretically investigated the performance of a new concept of near-infrared Fabry–Pérot graphene/silicon Schottky photodetector based on a double silicon on insulator substrate. The absorption mechanism, based on the internal photoemission effect, can be enhanced by exploiting the interference phenomena inside the optical microcavity. All numerical simulations have been carried out by the transfer matrix method and taking into account the physics behind the hot carrier emission from two-dimensional materials (graphene) into silicon. Moreover, for more accurate investigation, dispersion of all materials involved in the proposed structure have been taken into account, too.

We have investigated and compared two configurations: one where the current is collected in the transverse direction with respect to the direction of the incident light, the other where it is collected in the longitudinal direction. We prove that while the TCC is characterized by the highest graphene absorption, highest responsivity and lowest NEP, the LCC is characterized by the highest bandwidth and responsivity × bandwidth product. Our results show responsivity of 0.24 A/W, bandwidth in the GHz regime and noise equivalent power of 0.6 nW/cm$\sqrt{\text{Hz}}$. In addition, the devices show a spectral selectivity which could be tuned with a proper choice of the cavity thickness. In this work TCC is characterized by a best selectivity of 8.5 nm (FWHM) around 1550 nm.

A further increase in selectivity could be obtained by taking advantage of resonant structures characterized by higher-quality factors, moreover, thanks to the graphene broadband optical absorption these devices show the potentialities to work also at different wavelengths by simply changing the length of the three-layer cavity. The whole structure has been conceived to be compatible with silicon technology and we believe that it could have a huge impact in the field of silicon photonics. Of course, for a full CMOS compatibility some challenges need to be first addressed, among them: the transferring of large-area graphene preserving a reasonable mobility, the low-resistance interconnection with graphene during the back-end-of-line (BEOL) process and the choice of suitable dielectric and encapsulation schemes for hysteresis-free and low-voltage operations.

Funding: This research has received funding from the ATTRACT project funded by the EC under Grant Agreement 777222.

Conflicts of Interest: The author declares no conflict of interest.

References

1. Yole Dèvelop. Available online: https://www.slideshare.net/Yole_Developpement/silicon-photonics-2018-report-by-yole-developpement-85857212 (accessed on 2 June 2019).
2. Terracciano, M.; De Stefano, L.; Borbone, N.; Politi, J.; Oliviero, G.; Nici, F.; Casalino, M.; Piccialli, G.; Dardano, P.; Varra, M. Solid phase synthesis of a thrombin binding aptamer on macroporous silica for label free optical quantification of thrombin. *RCS Adv.* **2016**, *6*, 86762–86769. [CrossRef]
3. Managò, S.; Zito, G.; Rogato, A.; Casalino, M.; Esposito, E.; De Luca, A.C.; De Tommasi, E. Bioderived Three-Dimensional Hierarchical Nanostructures as Efficient Surface-Enhanced Raman Scattering Substrates for Cell Membrane Probing. *ACS Appl. Mater. Interfaces* **2018**, *10*, 12406–12416. [CrossRef] [PubMed]
4. Ishikawa, Y.; Wada, K. Germanium for silicon photonics. *Thin Solid Films* **2010**, *518*, S83–S87. [CrossRef]
5. Harame, D.L.; Koester, S.J.; Freeman, G.; Cottrel, P.; Rim, K.; Dehlinger, G.; Ahlgren, D.; Dunn, J.S.; Greenberg, D.; Joseph, A. The revolution in SiGe: Impact on device electronics. *Appl. Surf. Sci.* **2004**, *224*, 9–17. [CrossRef]
6. Koester, S.J.; Schaub, J.D.; Dehlinger, G.; Chu, J.O. Germanium-on-SOI infrared detectors for integrated photonic applications. *IEEE J. Sel. Top. Quantum Electron.* **2006**, *12*, 1489–1502. [CrossRef]
7. Larsen, A.N. Epitaxial growth of Ge and SiGe on Si substrates. *Mater. Sci. Semicond. Process.* **2006**, *9*, 454–459. [CrossRef]
8. Wang, J.; Lee, S. Ge-Photodetectors for Si-Based Optoelectronic Integration. *Sensors* **2011**, *11*, 696–718. [CrossRef]

9.   Casalino, M. Internal photoemission theory: Comments and theoretical limitations on the performance of near-infrared silicon schottky photodetectors. *IEEE J. Quantum Electron.* **2016**, *52*, 1–10. [CrossRef]

10.   Scales, C.; Berini, P. Thin-Film Schottky Barrier Photodetector Models. *IEEE J. Quantum Electron.* **2010**, *46*, 633–643. [CrossRef]

11.   Elabd, H.; Villani, T.; Kosonocky, W.F. Palladium-Silicide Schottky-Barrier IR-CCD for SWIR Applications at Intermediate Temperatures. *IEEE-ED Lett.* **1982**, *3*, 89–90. [CrossRef]

12.   Elabd, H.; Villani, T.S.; Tower, J.R. High Density Schottky-Barrier IRCCD Sensors for SWIR Applications at Intermediated Temperature. In Proceedings of the SPIE's Technical Symposium East, Arlington, VA, USA, 3–7 May 1982.

13.   Kosonocky, W.F.; Elabd, H.; Erhardt, H.G.; Shallcross, F.V.; Villani, T.; Meray, G.; Cantella, M.J.; Klein, J.; Roberts, N. 64 × 128-elements High-Performance PtSi IR-CCD Image Sensor. In Proceedings of the 1981 International Electron Devices Meeting, Washington, DC, USA, 7–9 December 1981.

14.   Kosonocky, W.F.; Elabd, H.; Erhardt, H.G.; Shallcross, F.V.; Meray, G.M.; Villani, T.S.; Groppe, J.V.; Miller, R.; Frantz, V.L.; Cantella, M.J. Design and Performance of 64 × 128-Element PtSi Schottky-Barrier IR-CCD Focal Plane Array. In Proceedings of the SPIE's Technical Symposium East, Arlington, VA, USA, 3–7 May 1982.

15.   Wang, W.L.; Winzenread, R.; Nguyen, B.; Murrin, J.J. High fill factor 512 × 512 PtSi focal plane array. In Proceedings of the SPIE's 33rd Annual Technical Symposium, San Diego, CA, USA, 22 December 1989.

16.   Casalino, M.; Sirleto, L.; Moretti, L.; Della Corte, F.; Rendina, I. Design of a silicon resonant cavity enhanced photodetector based on the internal photoemission effect at 1.55 µm. *J. Opt. A Pure Appl. Opt.* **2006**, *8*, 909–913. [CrossRef]

17.   Zhu, S.; Chu, H.S.; Lo, G.Q.; Bai, P.; Kwong, D.L. Waveguide-integrated near-infrared detector with self-assembled metal silicide nanoparticles embedded in a silicon p-n junction. *Appl. Phys. Lett.* **2012**, *100*, 061109. [CrossRef]

18.   Berini, P.; Olivieri, A.; Chen, C. Thin Au surface plasmon waveguide Schottky detectors on p-Si. *Nanotechnology* **2012**, *23*, 444011. [CrossRef] [PubMed]

19.   Akbari, A.; Tait, R.N.; Berini, P. Surface plasmon waveguide Schottky detector. *Opt. Express* **2010**, *18*, 8505–8514. [CrossRef] [PubMed]

20.   Knight, M.W.; Sobhani, H.; Nordlander, P.; Halas, N.J. Photodetection with active optical antennas. *Science* **2011**, *332*, 702–704. [CrossRef]

21.   Sobhani, A.; Knight, M.W.; Wang, Y.; Zheng, B.; King, N.S.; Brown, L.V.; Fang, Z.; Nordlander, P.; Halas, N.J. Narrowband photodetection in the near-infrared with a plasmon-induced hot electron device. *Nat. Commun.* **2013**, *4*, 1–6. [CrossRef]

22.   Desiatov, B.; Goykhman, I.; Mazurski, N.; Shappir, J.; Khurgin, J.B.; Levy, U. Plasmonic enhanced silicon pyramids for internal photoemission Schottky detectors in the near-infrared regime. *Optica* **2015**, *2*, 335–338. [CrossRef]

23.   Elabd, H.; Kosonocky, W.F. Theory and measurements of photoresponse of thin film Pd2Si and PtSi Schottky-barrier detectors with optical cavity. *RCA Rev.* **1982**, *43*, 569.

24.   Vickers, V.E. Model of Schottky barrier hot-electron-mode photodetection. *Appl. Opt.* **1971**, *10*, 2190–2192. [CrossRef]

25.   Di Bartolomeo, A.; Luongo, G.; Giubileo, F.; Funicello, N.; Niu, G.; Schroeder, T.; Lisker, M.; Lupina, G. Hybrid graphene/silicon Schottky photodiode with intrinsic gating effect. *2D Mater.* **2017**, *4*, 025075. [CrossRef]

26.   Luongo, G.; Grillo, A.; Giubileo, F.; Iemmo, L.; Lukosius, M.; Alvarado Chavarin, C.; Wenger, C.; Di Bartolomeo, A. Graphene Schottky Junction on Pillar Patterned Silicon Substrate. *Nanomaterials* **2019**, *9*, 659. [CrossRef] [PubMed]

27.   Levy, U.; Grajower, M.; Goncalves, P.A.D.; Mortensen, N.A.; Khurgin, J.B. Plasmonic silicon Schottky photodetectors: The physics behind graphene enhanced internal photoemission. *APL Photonics* **2017**, *2*, 026103. [CrossRef]

28.   Luongo, G.; Giubileo, F.; Genovese, L.; Iemmo, L.; Martucciello, N.; Di Bartolomeo, A. I-V and C-V Characterization of a High-Responsivity Graphene/Silicon Photodiode with Embedded MOS Capacitor. *Nanomaterials* **2017**, *7*, 158. [CrossRef] [PubMed]

29. Echtermeyer, T.J.; Britnell, L.; Jasnos, P.K.; Lombardo, A.; Gorbachev, R.V.; Grigorenko, A.N.; Geim, A.K.; Ferrari, A.C.; Novoselov, K.S. Strong plasmonic enhancement of photovoltage in graphene. *Nat. Commun.* **2011**, *2*, 458. [CrossRef] [PubMed]

30. Thongrattanasiri, S.; Koppens, F.H.L.; García de Abajo, F.J. Complete Optical Absorption in Periodically Patterned Graphene. *Phys. Rev. Lett.* **2012**, *108*, 047401. [CrossRef]

31. Konstantatos, G.; Badioli, M.; Gaudreau, L.; Osmond, J.; Bernechea, M.; Garcia de Arquer, P.; Gatti, F.; Koppens, F.H. Hybrid graphene-quantum dot phototransistors with ultrahigh gain. *Nat. Nanotechnol.* **2012**, *7*, 363–368. [CrossRef]

32. Unlu, M.S.; Strite, S. Resonant cavity enhanced photonic devices. *J. Appl. Phys.* **1995**, *78*, 607–639. [CrossRef]

33. Furchi, M.; Urich, A.; Pospichil, A.; Lilley, G.; Unterrainer, K.; Detz, H.; Klang, P.; Andrews, A.M.; Schrenk, W.; Strasse, G.; et al. Microcavity-Integrated Graphene Photodetector. *Nanoletters* **2012**, *12*, 2773–2777. [CrossRef]

34. Emsley, M.K.; Dosunmu, O.; Unlu, M.S. High-Speed Resonant-Cavity-Enhanced Silicon Photodetectors on Reflecting Silicon-On-Insulator Substrates. *IEEE Photonics Technol. Lett.* **2002**, *14*, 519–521. [CrossRef]

35. Casalino, M.; Russo, R.; Russo, C.; Ciajolo, A.; Di Gennaro, E.; Iodice, M. Free-Space Schottky Graphene/Silicon Photodetectors operating at 2 μm. *ACS Photonics* **2018**, *5*, 4577–4585. [CrossRef]

36. Tongay, S.; Lemaitre, M.; Miao, X.; Gila, B.; Appleton, B.R.; Hebard, A.F. Rectification at Graphene-Semiconductor Interfaces: Zero-Gap Semiconductor-Based Diodes. *Phys. Rev. X* **2012**, *2*, 011002. [CrossRef]

37. Sze, S.M.; Ng, K.K. *Physics of Semiconductor Devices*; John Wiley & Sons: Hoboken, NJ, USA, 2006.

38. Unlu, S.M.; Onat, B.M.; Leblebici, Y. Transient simulation of heterojunction photodiode-part II: Analysis of resonant cavity enhanced photodetectors. *J. Lightwave Technol.* **1995**, *13*, 406–415. [CrossRef]

39. BEA, S.; Teich, M.C. *Fundamentals of Photonics*; John Wiley & Sons: Hoboken, NJ, USA, 1991.

40. Casalino, M. Design of Resonant Cavity-Enhanced Schottky Graphene/Silicon Photodetectors at 1550 nm. *J. Ligthwave Technol.* **2018**, *36*, 1766–1774. [CrossRef]

41. Palik, E.D. *Handbook of Optical Constants of Solids*; Academic Press: San Diego, CA, USA, 1998.

42. Casalino, M.; Coppola, G.; Gioffrè, M.; Iodice, M.; Moretti, L.; Rendina, I.; Sirleto, L. Cavity Enhanced Internal Photoemission Effect in Silicon Photodiode for Sub-Bandgap Detection. *J. Ligthwave Technol.* **2010**, *28*, 3266–3272. [CrossRef]

43. RefractiveIndex.INFO. Available online: https://refractiveindex.info (accessed on 2 June 2020).

44. Echtermeyer, T.J.; Milana, S.; Sassi, U.; Eiden, A.; Wu, M.; Lidorikis, E.; Ferrari, A.C. Surface Plasmon Polariton Graphene Photodetectors. *Nano Lett.* **2016**, *16*, 8–20. [CrossRef]

45. Nair, R.R.; Blake, P.; Grigorenko, A.N.; Novoselov, K.S.; Booth, T.J.; Stauber, T.; Peres, N.M.R.; Geim, A.K. Fine Structure Constant Defines Visual Transparency of Graphene. *Science* **2008**, *320*, 1308. [CrossRef]

46. Pavesi, L.; Guillot, G. *Optical Interconnect*; Springer-Verlag Berlin Heidelberg: Berlin, Germany, 2006; p. 64.

47. Lupina, G.; Strobel, C.; Dabrowski, J.; Lippert, G.; Kitzmann, J.; Krause, H.M.; Wenger, C.; Lukosius, M.; Wolff, A.; Albert, M. Plasma-enhanced chemical vapor deposition of amorphous Si on graphene. *Appl. Phys. Lett.* **2016**, *108*, 193105. [CrossRef]

48. Emsley, M.K.; Dosunmu, O.; Unlu, M.S. Silicon substrates with buried distributed Bragg reflectors for resonant cavity-enhanced optoelectronics. *IEEE J. Sel. Top. Quantum Electron.* **2002**, *8*, 948–955. [CrossRef]

49. Muriel, M.A.; Carballar, A. Internal field distributions in fiber Bragg gratings. *IEEE Photonics Technol. Lett.* **1997**, *9*, 955–957. [CrossRef]

50. Casalino, M.; Sassi, U.; Goykhman, I.; Eiden, A.; Lidorikis, E.; Milana, S.; De Fazio, D.; Tomarchio, F.; Iodice, M.; Coppola, G.; et al. Vertically Illuminated, Resonant Cavity Enhanced, Graphene-Silicon Schottky Photodetectors. *ACS Nano* **2017**, *11*, 10955–10963. [CrossRef]

Article

Effect of Various Defects on 4H-SiC Schottky Diode Performance and Its Relation to Epitaxial Growth Conditions

Jinlan Li [1,2], Chenxu Meng [2], Le Yu [2], Yun Li [3], Feng Yan [2], Ping Han [2,*] and Xiaoli Ji [2,*]

1   College of Information Engineering, Yangzhou University, Yangzhou 225009, China; 007328@yzu.edu.cn
2   College of Electronic Science and Engineering, Nanjing University, Nanjing 210093, China;
    mg1823040@smail.nju.edu.cn (C.M.); yule@nju.edu.cn (L.Y.); fyan@nju.edu.cn (F.Y.)
3   Science and Technology on Monolithic Integrated Circuits and Modules Laboratory, Electronic Devices
    Institute, Nanjing 210016, China; yuefei_2@126.com
*   Correspondence: hanping@nju.edu.cn (P.H.); xji@nju.edu.cn (X.J.);
    Tel.: +86-025-83685327 (P.H.); +86-025-89683965 (X.J.)

Received: 29 May 2020; Accepted: 22 June 2020; Published: 24 June 2020

Abstract: In this paper, the chemical vapor deposition (CVD) processing for 4H-SiC epilayer is investigated with particular emphasis on the defects and the noise properties. It is experimentally found that the process parameters of C/Si ratio strongly affect the surface roughness of epilayers and the density of triangular defects (TDs), while no direct correlation between the C/Si ratio and the deep level defect $Z_{1/2}$ could be confirmed. By adjusting the C/Si ratio, a decrease of several orders of magnitudes in the noise level for the 4H-SiC Schottky barrier diodes (SBDs) could be achieved attributing to the improved epilayer quality with low TD density and low surface roughness. The work should provide a helpful clue for further improving the device performance of both the 4H-SiC SBDs and the Schottky barrier ultraviolet photodetectors fabricated on commercial 4H-SiC wafers.

Keywords: Ni/4H-SiC Schottky barrier diodes (SBDs); C/Si ratios; 1/f noise

1. Introduction

As an excellent wide bandgap semiconductor material, 4H-SiC has attracted continuous attention in the last several decades. Due to its wide bandgap, high thermal conductivity, high saturated electron drift velocity and high physical and chemical stability, 4H-SiC is an ideal material for high-performance ultraviolet (UV) photodetectors available in high-temperature environments, as well as power devices for high-temperature and high-frequency applications [1–6]. The device performance of both the photodetectors and the power devices greatly depends on the epitaxial quality of commercial 4H-SiC wafers, which has been significantly improved in recent years. However, due to substrate material imperfection and epitaxial growth immaturity, there are still various defects in the large-size 4H-SiC epilayers, such as morphological defects (micropipes, downfalls, particles, triangular defects (TDs), carrots, etc.), crystallographic defects (stacking faults, threading dislocations, etc.) and deep level defects, etc. [7–15], which will inevitably have a serious impact on the performance of 4H-SiC power devices and ultraviolet photodetectors. Among the morphological defects, micropipes have serious impact on the device performance but are preventable, since the substrates with micropipe density less than 0.1 cm^{-2} are commercially available [16–18]. In our previous paper, we reported the density of other morphological defects in a 4-inch 4H-SiC epitaxial wafer, as well as their probability of causing Schottky diode power device failure [19]. The average morphological defect (including TDs, downfalls, particles, carrots) density in the 4-inch 4H-SiC epitaxial layers is 1.25 cm^{-2}, in which the TDs, the downfalls, the particles, and the carrots account for approximately 60%, 2%, 31% and 7%, respectively. In addition, the probability of causing device failure is 100% for TDs and downfalls, while the probability of degrading the reverse breakdown voltage of the device is 2% and 30% for particles and carrots, respectively. Therefore, among all these morphological defects, the TDs should attract enough attention as the critical device killer. Meanwhile, the deep level defect $Z_{1/2}$ center (located at 0.6~0.7 eV below the conduction band of 4H-SiC) is now considered to be the main life killer of n-type 4H-SiC devices. The concentration of the $Z_{1/2}$ is $(0.1–4) \times 10^{13}$ cm^{-3} for the commercially available 4H-SiC material, which seriously affects the performance of bipolar devices and photodetectors due to the existence of deep level traps [20]. Thus, the TDs and the $Z_{1/2}$ defects which widely exist in commercially available 4H-SiC epilayers are usually considered as two significant factors that hinder the device performance. The structures of TDs are related to the presence of foreign particle inclusion at the epi-substrate interface, resulting in an increase of leakage current and the reduction of breakdown voltage in the SiC power devices [21]. The $Z_{1/2}$ defect at E_c-0.6 eV originates from carbon vacancies, which may affect the barrier height and leakage current of SiC Schottky barrier diodes (SBDs) [22]. On the other hand, all these defects may increase the recombination of electron-hole pairs and enhance the surface recombination in the Schottky barrier UV photodetectors, leading to a lower responsivity of the device [23,24]. A large number of studies have shown that the formation of these defects is closely associated with growth temperature, C/Si ratio, thermal oxidation, carbon ion implantation, thermal annealing, in-situ pre-growth etch and cutting angle of SiC substrate, etc. [25–30].

Therefore, the optimization of growth conditions is critical to the reduction of defects and the improvement of the device performance. Miyazawa reported that the reduction of the $Z_{1/2}$ defect concentration is achieved mainly through thermal oxidation/thermal annealing or C^+-implantation/thermal annealing, unfortunately, new traps are introduced by these two methods [31]. Yazdanfar pointed out that TD density can be effectively reduced by in-situ pre-growth etch before growth at lower C/Si ratio and growth temperature [32]. However, these methods are very complicated and impractical for industrial processes. In practicality, the variations of growth conditions influence multiple defects in the 4H-SiC epilayers, and consequently determine the device performance in an intricate way. It is necessary to survey how these parameters (e.g., C/Si ratio, growth rate and doping density) affect TDs and $Z_{1/2}$ defects, which are the most commonly exsiting defects in the commercial 4H-SiC wafers. Meanwhile, the key issue of process parameters optimization is to seek out the critical factors affecting device performance; therefore, it is also important to establish a direct link between the growth parameters and the device performance. By analyzing the results of the electrical parameters of the device performance, not only can the most important defect information affecting the performance be identified, but it can also provide effective suggestions for further optimization of the growth parameters.

In this paper, we investigated the influence of different chemical vapor deposition (CVD) growth conditions on the defect density of 4H-SiC epilayers and the electrical properties of Ni/4H-SiC SBDs, with particular attention to the TDs and the $Z_{1/2}$ defects. The integration of scanning electron microscopy (SEM), atomic force microscopy (AFM), micro-photoluminescence (PL), and micro-Raman was used to study the mechanism of defect evolution in 4-inch 4H-SiC epilayers. The combination of reverse leakage current testing, noise measurement, and the deep level transient spectrum testing (DLTS) was used to reveal the root cause of the growth conditions impacting on device performance.

2. Materials and Methods

4H-SiC epilayers with different C/Si ratios (C/Si = 0.9, 1 or 1.1) were homoepitaxially grown on commercially available 4H-SiC substrates in a SiH_4-C_3H_8-H_2 chemical vapor deposition (CVD) system (AIXTRON VP2400, AIXTRON, Herzogenrath, Germany). In this study, all substrates were commercial chemo-mechanically polished micropipe-free 4-inch 4H-SiC. The substrate is highly doped with nitrogen and cut off at 8° towards [11$\bar{2}$0] direction. Here, SiH_4, C_3H_8 and HCl were used as source gases while H_2 was selected for dilution and carrier gas. The C/Si ratio was varied from 0.9 to 1.1 by changing the C_3H_8 flow rate at a fixed SiH_4 flow rate. The typical epitaxial growth rates were fixed at 60 μm/h, and the epitaxial temperature and growth pressure were controlled within

1550~1600 °C and 50~150 mbar, respectively. Nitrogen gas was used as an n-type dopant, and the typical doping concentration was 1×10^{15} cm^{-3}. Additional 4-inch 4H-SiC epilayers with different growth rate ($v = 30$ μm/h and 60 μm/h) and doping concentration ($N_d = 4 \times 10^{15}$ cm^{-3}, 7.5×10^{15} cm^{-3} and 1×10^{16} cm^{-3}) were also prepared for exploring the dependence of the process parameters on deep defect density, respectively. All the samples (#1~#7) have the same thickness of 12 μm, with different growth parameters summarized in Table 1. After the homoepitaxial growth in CVD system, a 100 nm Ni Ohmic contact was formed by annealing in nitrogen at 1000 °C for 5 min in the back sides of the device, and a 75 nm Ni Schottky contact layer was sputtered with an active area of 1 mm^2 in the front sides of the device.

Table 1. The samples of the 4-inch 4H-SiC epilayers with different chemical vapor deposition (CVD) growth parameters.

Samples #	#1	#2	#3	#4	#5	#6	#7
C/Si	0.9	1	1.1	0.8	0.8	0.8	0.8
Growth rate (μm/h)	60	60	60	60	60	60	30
Doping density (10^{15} cm^{-3})	1	1	1	4	7.5	10	7.5

The room temperature micro-PL and micro-Raman spectroscopy measurement were performed on the samples excited by a 325 nm He-Cd laser, where the laser beam was focused to a spot of 10 μm diameter using a sapphire objective lens. Moreover, the atomic force microscopy (AFM) (WET-SPM-model, SPM-9600, Shimadzu Corp., Kyoto, Japan) imaging with a metalized cantilever was carried out for characterizing the surface roughness of wafers. The *C-V* characteristic measurements (Keithley 4200, Keithley, Cleveland, OH, USA) were conducted at a frequency of 1 MHz to determine the effective doping density of devices based on $1/C^2$ vs. V plots. The noise spectrum was tested to study device noise performance. Deep level characterization of the 4H-SiC SBDs were carried out using DLS-83D Deep Level Transient Spectroscopy test system (Semilab, Budapest, Hungary). Temperature ranging from 77 K to 550 K was selected by a ACP-4000 temperature controller at a heating rate of 0.1 K/s.

3. Results and Discussion

Figure 1a shows the room temperature micro-PL spectra corresponding to the A and B positions in the illustration for the 4H-SiC epilayer grown with C/Si = 1.1. The inset SEM image at position A exhibits the crystal structure region of the 4H-SiC epilayer without TDs. The strongest peak of 391 nm corresponding to 3.17 eV is attributed to the typical band edge emission of 4H-SiC [33]. However, at the apex of the TDs (position B), the 4H-SiC band edge emission intensity at 391 nm was significantly reduced. Furthermore, an additional small emission peak appeared at 423 nm corresponding to 2.93 eV, which is consistent with the emission wavelength of the stacking fault in the 4H-SiC epilayer. This stacking fault is determined by the different stacking order, which is defined as a single Shockley SF (1SSF) using the Zhdanov notation [34,35]. Figure 1b shows typical micro-Raman spectra of 4H-SiC at points A and B. The results show two FTO modes at 776 cm^{-1} and 796 cm^{-1}, and an FLO mode at 964 cm^{-1}, which is consistent with the typical Raman spectra of 4H-SiC [36]. However, for the TD region (point B), the intensity of the Raman peak at 796 cm^{-1} is gradually increased. Since the 3C-SiC also has a TO peak at 796 cm^{-1}, the presence of the 3C-SiC polytype could be inferred. The ratio of the two peaks can be taken as a measure of the presence of the 3C-SiC polytype [37]. Therefore, we can confirm that stacking faults and the 3C-SiC are the reasons of nucleation at the triangle defect vertex. Figure 2 shows the dependence of TD density on the C/Si ratios for 4H-SiC epilayers. The wafers (#1~#3) were inspected using optical microscopy to evaluate the density of TDs. The results show that the density of TDs is reduced from 1.3 cm^{-2} to 0.13 cm^{-2} with the C/Si ratio decreasing from 1.1 to 0.9. Kojima et al. have reported that the origin of the TDs is attributed to step-bunching caused by the interrupted step-flow growth [38]. In the carbon-rich growth environment (C/Si = 1.1), the surface free energy of the crystal surface is higher than that under silicon-rich conditions, so it is necessary to reduce the surface free energy by the formation of step-bunching [39]. Thus, there is a higher probability of two-dimensional nucleation due to the suppression of the step-flow growth at high C/Si ratios. Consequently, a relatively low C/Si ratio can effectively reduce the TD density.

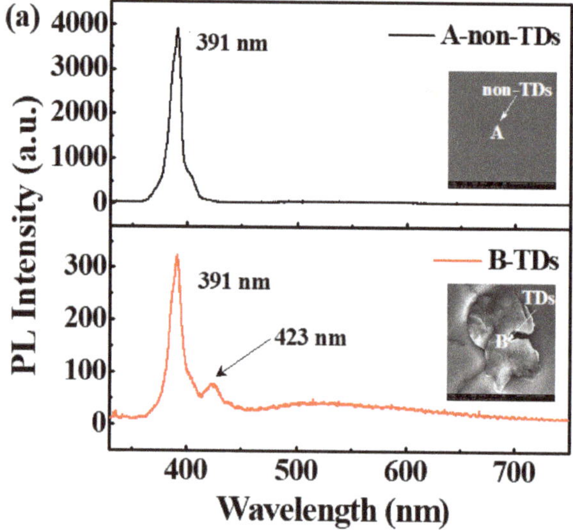

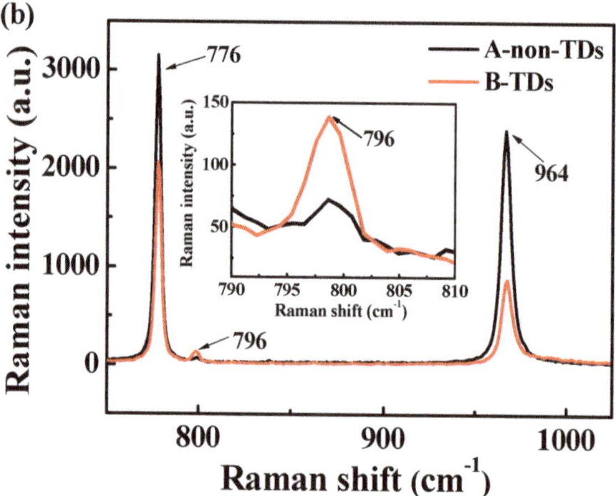

Figure 1. (**a**) Room temperature micro-photoluminescence (PL) spectra. The insets show scanning electron microscopy (SEM) images of region A (non-triangular defects (TDs)) and region B (TDs); (**b**) Micro-Raman spectra corresponding to the A and B positions in the 4H-SiC epilayer grown with C/Si = 1.1. The inset is a comparison of the intensity of the Raman peak at 796 cm^{-1} in the enlarged view.

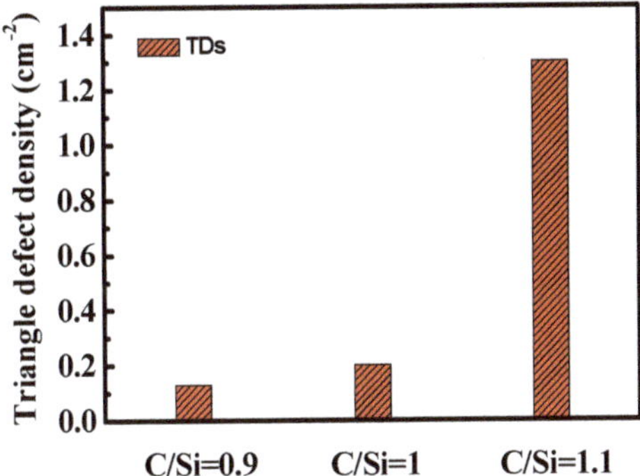

Figure 2. The C/Si ratio dependence of TDs density for 4H-SiC epilayers (#1~#3).

It is well known that TDs are fatal to power devices, directly causing device failure. Moreover, although the TDs could be got rid of in most areas of the epitaxial wafer by a proper C/Si ratio, there still exist some hidden defects that affect device performance, such as interface states and deep defects, etc. Therefore, we selected the devices without TDs on the epilayer region to further investigate the effect of hidden defects on the performance of 4H-SiC SBDs prepared on epilayers with different C/Si ratios (C/Si = 0.9, 1 or 1.1). Figure 3a shows the mean values of the reverse current density varying with the C/Si ratios under the reverse bias voltage of −200 V. It is noted that there is still a minor difference in device performance due to process unevenness even in the same wafer. In order to eliminate the influence of the performance non-uniformity between devices, we analyzed the average reverse *I-V* results of five samples randomly selected on each wafer, and Ni/4H-SiC SBDs with C/Si = 1 is shown as an example in the inset of Figure 3a. It is found that the average values of the reverse current density of Ni/4H-SiC SBDs with C/Si = 0.9, 1 or 1.1 are 1.8×10^{-12} A/cm^2, 5.6×10^{-12} A/cm^2, and 1.5×10^{-11} A/cm^2, respectively. With the increasing of C/Si ratios, the reverse leakage current gradually increases. Compared with the sample with C/Si = 0.9, the leakage current of the sample with C/Si = 1.1 increases by nearly an order of magnitude. Figure 3b shows the forward *I-V* characteristics of Ni/4H-SiC SBDs under different C/Si ratios. It is seen that all samples have relatively uniform and consistent forward *I-V* curves. According to the thermionic emission (TE) theory, the relationship between current and voltage is defined by [40]

$$J = A^*T^2\exp(\frac{-q\Phi_B}{kT}) \left[\exp\frac{qV}{nkT} - 1\right] \tag{1}$$

Here, the values of the barrier height Φ_B and the ideality factor n can be calculated by

$$n = \frac{q}{kT}\left(\frac{dV}{d\ln J}\right) \tag{2}$$

$$\Phi_B = \frac{kT}{q}\ln\left(\frac{A^*T^2}{J_s}\right) \tag{3}$$

where n is the ideality factor, V is the applied voltage, Φ_B is the zero bias Schottky barrier height and A^* is the effective Richardson constant with 146 A/cm^{2}K^2 for 4H-SiC. The values of ideality factor n and barrier height Φ_B obtained from *I-V* characteristics, and the values of effective doping density N_{eff} extracted from *C-V* characteristics in the inset of Figure 3b were summarized in Table 2. The n and Φ_B are all approximately 1 and 1.63 eV, showing good rectification characteristics. Moreover, the N_{eff} are in the range of $1.15\sim1.25 \times 10^{15}$ cm^{-3}. Therefore, the barrier height and the effective doping density are not the root cause of the increase in reverse leakage current as the C/Si ratios increases, because of the same barrier height for all samples. As is known, the leakage current of devices may be affected by a lot of factors, including surface defects, crystal defects, deep level defects, interface states, and crystal quality, etc. [41–43]. Combined with the above analysis, we can infer that the epitaxy process with different C/Si ratios may affect the interface state, deep level defects or crystal quality of the epilayer, resulting in different reverse leakage current changing with C/Si ratios.

Table 2. The electrical parameters of 4H-SiC SBDs with different C/Si ratios (C/Si = 0.9, 1 or 1.1).

Samples #	#1	#2	#3
n	1.008	1.004	1.010
Φ_B (eV)	1.629	1.631	1.629
N_{eff} (10^{15} cm^{-3})	1.25	1.17	1.15

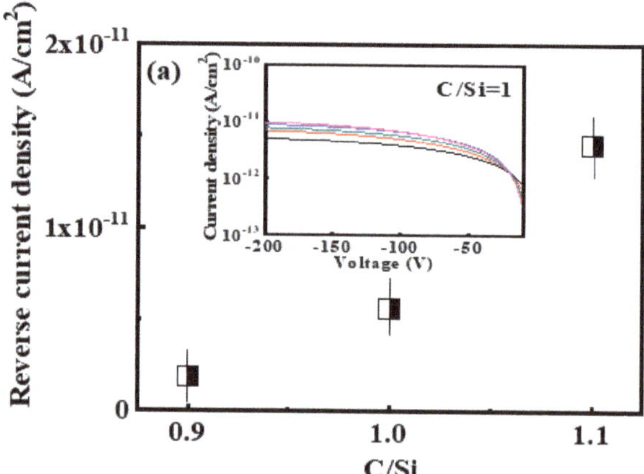

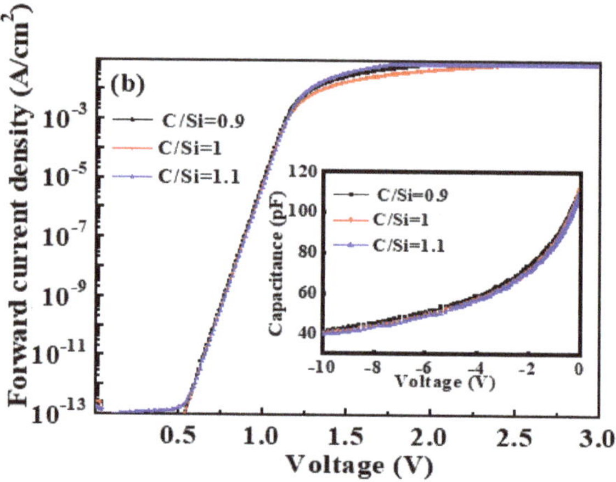

Figure 3. *Cont.*

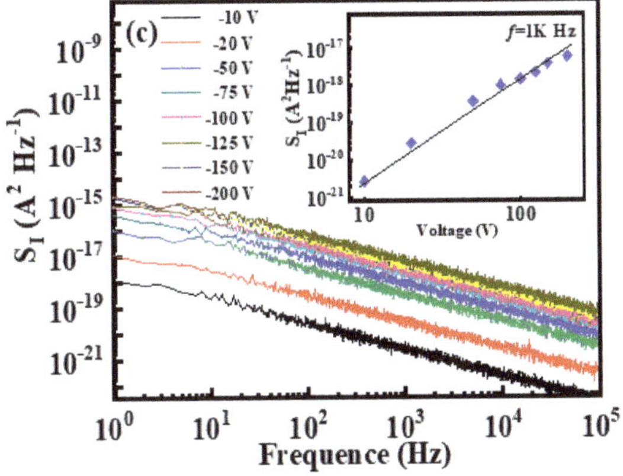

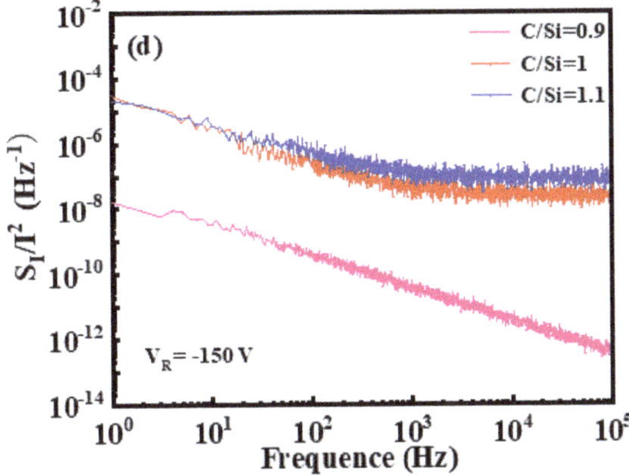

Figure 3. (**a**) The dependence of the mean reverse current density on C/Si ratio for Ni/4H-SiC Schottky barrier diodes (SBDs) under $V_R = -200$ V. The inset shows reverse *I-V* characteristics of a representative sample (Ni/4H-SiC SBDs with C/Si = 1); (**b**) Forward *I-V* characteristics of Ni/4H-SiC SBDs under different C/Si ratios (C/Si = 0.9, 1 or 1.1). The inset shows the *C-V* characteristics of Ni/4H-SiC SBDs; (**c**) Frequency dependence of the spectral noise density (S_I) for Ni/4H-SiC SBDs with C/Si = 0.9 at room temperature under $V_R = -10 \sim -200$ V, the inset shows the bias voltage dependence of the spectral noise density at 1K Hz; (**d**) Noise spectra of Ni/4H-SiC SBDs with C/Si ratios (C/Si = 0.9, 1 or 1.1).

Figure 3c show the frequency and bias voltages dependence of the spectral noise density (S_I) for the Ni/4H-SiC SBDs with C/Si=0.9 at room temperature under various reverse bias voltages ($V_R = -10 \sim (-200)$ V). It is clearly observed that the spectral noise density is inversely proportional to the frequency at different voltages and increases approximately linearly with the increase of the reverse bias voltage. It has the form of $1/f^\alpha$ noise, where $\alpha=1$, its characteristics is flicker noise ($1/f$ noise). This dependence is consistent with the reported results of SiC Schottky diodes ($\alpha = 0.5 \sim 1.5$) [44,45]. The main contribution to $1/f$ noise and resistance noise comes from the Schottky barrier. Figure 3d shows the noise spectra of the Ni/4H-SiC SBDs with varying C/Si ratios (C/Si = 0.9, 1 or 1.1) under $V_R = -150$ V. It displays $1/f$ behavior at frequencies below 1K Hz for all samples. Furthermore, the spectral noise density increases with the increase of the C/Si ratio, and the minimum is obtained at C/Si = 0.9. However, for the samples with C/Si = 1 and 1.1, the noise spectra become almost frequency independent at frequencies greater than 1K Hz. Several orders of magnitude increase in the noise are caused by superimposed thermal noise ($S_I = 4K_0Tg$) due to larger reverse current in samples with C/Si = 1 or 1.1, which is in agreement with the results of Figure 3a. The $1/f$ noise is mainly affected by the factors such as the interface state and crystal quality of the semiconductor material. Zhang et al. reported that the $1/f$ noise is reduced significantly with the increase of temperature in the SiC MOSFET within a temperature range of 85~510 K, which is attributed to the interface trap density decreasing with the temperature [46]. Soibel et al. also revealed that the $1/f$ noise in the InAs/GaSb superlattice detector is related to the side leakage current caused by the surface states. The larger side leakage current is induced by the higher surface state density, leading to a significant increase in noise [47]. Accordingly, our results show that as the C/Si ratio decreases, the spectral noise density decreases by about three orders of magnitude (C/Si = 0.9). The influence of C/Si ratios on device noise should be due to the surface state density or the interface state density caused by different crystal quality of 4H-SiC epilayers. Therefore, an optimized C/Si ratio not only effectively improves the performance of SBDs, but should also improves the responsivity of the Schottky barrier UV photodetector due to the reduction of noise.

On the other hand, the $Z_{1/2}$ is the most important deep level defect in the grown N-type 4H-SiC epilayer. The presence of deep level defects acts as charge trapping, capturing photogenerated carriers or increasing their recombination probability in the depletion region of the Schottky barrier UV photodetectors, leading to a significant reduction of responsivity [48]. Moreover, the trap or defect-assisted tunneling can result in the increase of leakage current and the reduction of carrier lifetime. To further understand the effects of deep defects on the electrical properties of the Ni/4H-SiC SBDs with different growth parameters, DLTS measurements were carried out in a temperature range of 80~550 K. Figure 4a shows the representative DLTS spectra of the 4H-SiC SBD grown with a C/Si ratio of 0.9, under the reverse bias $V_R = -3$ V, pulse voltage $V_p = 0$ V, pulse width $t_p = 50$ μs and frequency $f = 80$~960 Hz. It can be seen that the only peak appears in the DLTS spectrum, which can be attributed to $Z_{1/2}$ as an intrinsic defect in the SiC epilayer. The microstructure of $Z_{1/2}$ has been extensively studied, such as carbon vacancies (V_c), silicon vacancies (V_{si}), reverse (C_{si}, Si_c) or more likely defect complexes ($C_{si} + Si_c$, $V_c + V_{si}$) [49]. Eberlein et al. have reported the $Z_{1/2}$ is composed of the traps Z_1 and Z_2, and such centers reveal the negative-U character with donor (0/+) levels located at E_c-0.43(0.46) eV and acceptor (−/0) levels situated at E_c-0.67(0.71) eV [50]. The $Z_{1/2}$ has recently been showned to be most likely an acceptor level of V_c [51].Besides, the variation of $Z_{1/2}$ defect concentration in all samples (#1~#7) is shown in Figure 4b. The parameters of defects are summarized in Table 3. The results show that with the change of C/Si ratio (C/Si = 0.9~1.1), the $Z_{1/2}$ defect concentration is 2.28×10^{13} cm^{-3}, 3.79×10^{13} cm^{-3}, and 2.22×10^{13} cm^{-3}, respectively, indicating that the $Z_{1/2}$ defect concentration remains substantially unchanged. Litton et al. reported that the $Z_{1/2}$ concentration decreased with increasing carbon to silicon ratios of one, three and six due to the reduction of C vacancies under a C-rich growth condition. However, as the carbon-to-silicon ratio increases, the TD density of its epitaxial wafers will increase sharply, which will directly cause the decrease of device performance. Compared to our experimental results, the $Z_{1/2}$ defect density does not change significantly with the change of the carbon-to-silicon ratio may be due to the small range of change of the carbon-to-silicon ratio [49]. In addition, as the doping concentration and growth rate increase, the $Z_{1/2}$ defect concentration ranges between 6.78×10^{12} cm^{-3} and 1.41×10^{13} cm^{-3}. Lilja et al. reported that the $Z_{1/2}$ concentration does not show any obvious tendency with epitaxial growth rate, and the concentration of $Z_{1/2}$ in different epitaxial wafers is between 1×10^{13} cm^{-3} and 4×10^{13} cm^{-3} [52]. The above literature results is in agreement with our results. Therefore, the above experimental results show that the concentration of $Z_{1/2}$ intrinsic defect does not change significantly with the process parameters. The inset of Figure 4b shows the dependence of the reverse current density (under $V_R = -200$ V) and breakdown voltage as a function of $Z_{1/2}$ defect density for Ni/4H-SiC SBDs. It can be seen from the figure that the reverse current and the breakdown voltage have no obvious trend with the increase of $Z_{1/2}$ defect density. However, as the carbon-to-silicon ratio decreases from 1.1 to 0.9, the reverse current gradually decreases, and the breakdown voltage slightly increases, which may be due to the optimization of interface quality leading to improved device performance. In addition, devices with different C/Si ratios exhibit uniform and good rectification characteristics. (the n and Φ_B are all approximately 1 and 1.63 eV, respectively). Therefore, the $Z_{1/2}$ defect is not the root cause of device performance degradation.

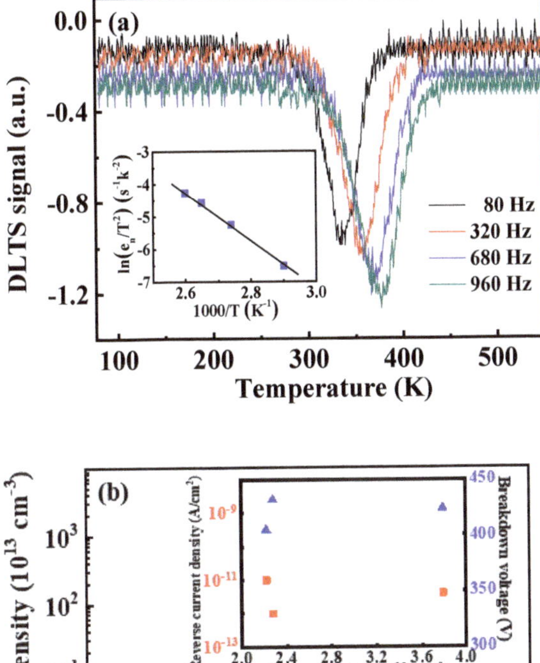

Figure 4. (**a**) Deep level transient spectrum testing (DLTS) spectra of Ni/4H-SiC SBD with C/Si = 0.9; (**b**) The $Z_{1/2}$ defect concentration of Ni/4H-SiC SBDs under different CVD growth conditions. The inset shows the dependence of the reverse current density under $V_R = -200$ V (red square symbols) and breakdown voltage (blue triangle symbols) on $Z_{1/2}$ defect concentration for Ni/4H-SiC SBDs with C/Si ratios (C/Si = 0.9, 1 or 1.1).

Table 3. The $Z_{1/2}$ defect parameters obtained from the DLTS measurements of the 4-inch 4H-SiC epilayers with different growth parameters.

Samples #	ΔE (eV)	σ (cm²)	N_t (cm⁻³)
#1	E_c-0.627	1.18×10^{-15}	2.28×10^{13}
#2	E_c-0.626	8.01×10^{-16}	3.79×10^{13}
#3	E_c-0.624	5.06×10^{-16}	2.22×10^{13}
#4	E_c-0.648	2.42×10^{-16}	7.62×10^{12}
#5	E_c-0.644	2.44×10^{-15}	6.78×10^{12}
#6	E_c-0.687	8.33×10^{-15}	1.24×10^{13}
#7	E_c-0.610	5.96×10^{-16}	1.41×10^{13}

In order to further study the effect of C/Si ratio on the reverse leakage and noise of the device, Figure 5a shows the AFM images of the 4H-SiC epilayers grown with different C/Si ratios from 1.1 to 0.9. It is exhibited that the strong dependence of surface morphology of 4H-SiC epilayers on the C/Si ratios. It is observed that the surface morphology is rather smooth with gentle undulation under Si-rich growth environment (C/Si = 0.9). The root-mean-square (RMS) roughness is 0.8 nm at a C/Si ratio of 1.1, and it is decreased to 0.15 nm at C/Si ratio of 0.9 in a 3 μm × 3 μm scan area. As seen in Figure 5b, the PL spectra of 4H-SiC epilayers with different C/Si ratios shows that the 4H-SiC band edge emission intensity (391 nm) of the sample with C/Si = 1.1 is significantly weaker than that of the sample with C/Si = 0.9, which is due to the fact that the non-radiative recombination caused by the defects severely weakens the band edge emission intensity of 4H-SiC [34]. Hence, we can see that the sample with C/Si = 0.9 has minimal roughness and defects, resulting in minimum noise. By comparison, in the samples of C/Si = 1 or 1.1, the quality of the wafer becomes worse, thus causing more noise.

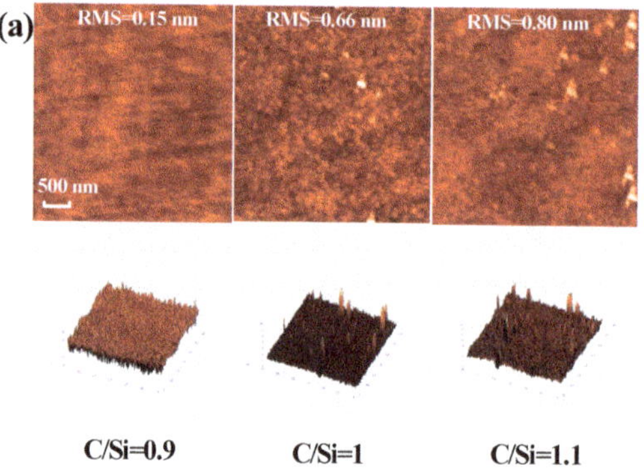

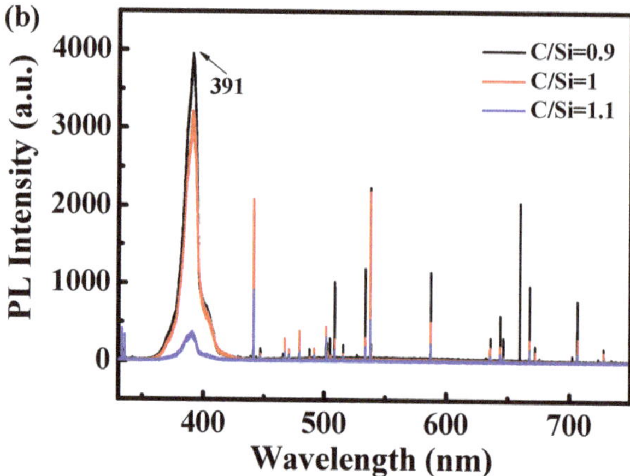

Figure 5. (**a**) Atomic force microscopy (AFM) images and (**b**) PL spectra of 4H-SiC epilayers with C/Si ratios (C/Si = 0.9, 1 or 1.1).

Combined with the above test result, we can see that the roughness of the epitaxial wafer and the crystal quality associated with the defects play an important role in device performance. Since the $Z_{1/2}$ defect density and barrier height of 4H-SiC SBDs do not change with the different CVD growth conditions, the most probable reason for the difference in reverse leakage and noise of the device should be the different interface state densities caused by the different epitaxial quality. Therefore, the optimization of the C/Si ratio not only effectively reduces the TD density but also improves the performance of 4H-SiC SBD.

4. Conclusions

The impacts of CVD growth parameters on defect density of 4-inch 4H-SiC epilayers and performance of Ni/4H-SiC SBDs were studied. It is found that the reverse current and noise characteristics of 4H-SiC SBDs are highly dependent on the C/Si ratio. As the C/Si ratio grows from 0.9 to 1.1, the average reverse current and $1/f$ noise of 4H-SiC SBDs increase gradually. Furthermore, the DLTS characterization clarifies that the $Z_{1/2}$ defect is not the root cause of the device performance because the $Z_{1/2}$ defect density is almost unchanged with the growth conditions. The AFM and PL tests further confirm that the crystal quality of the sample becomes worse with the increase of the C/Si ratio by comparing the RMS roughness and the SiC band edge emission intensity (391 nm). Therefore, it can be concluded that the increase of leakage current and noise are due to the crystal quality of 4H-SiC epilayers. The optimization of the C/Si ratio can significantly reduce TD density and improve the crystal quality of 4H-SiC epilayers, which further enhances the electrical properties of the 4H-SiC SBDs. Moreover, the study on the influence of C/Si ratio on TDs and $Z_{1/2}$ defects should also be helpful for improving the responsivity of the 4H-SiC Schottky barrier UV photodetectors.

Author Contributions: Conceptualization, X.J. and P.H.; data curation, X.J.; investigation, J.L.; visualization, J.L.; writing—original draft preparation, J.L.; writing—review and editing, X.J., L.Y., C.M., F.Y. and Y.L.; All authors have read and agreed to the published version of the manuscript.

Funding: This research was funded by National Key Research and Development Program of China, grant number 2016YFB0400402, 2016YFB0402403, 2016YFA0202102.

Conflicts of Interest: The authors declare no conflict of interest.

References

1. Hefner, A.R.; Singh, R.; Lai, J.S.; Berning, D.W.; Bouche, S.; Chapuy, C. SiC power diodes provide breakthrough performance for a wide range of applications. *IEEE Trans. Power Electron.* **2001**, *16*, 273–280. [CrossRef]
2. Licciardo, G.D.; Bellone, S.; Benedetto, L.D. Analytical model of the forward operation of 4H-SiC vertical DMOSFET in the safe operating temperature range. *IEEE Trans. Power Electron.* **2015**, *30*, 5800–5809. [CrossRef]
3. Wang, X.; Cooper, J.A. High-Voltage n-Channel IGBTs on Free-Standing 4H-SiC Epilayers. *IEEE Trans. Electron Devices* **2010**, *57*, 511–515. [CrossRef]
4. Ren, N.; Wang, J.; Sheng, K. Design and Experimental Study of 4H-SiC trenched junction barrier Schottky diodes. *ISO IEEE Trans. Electron Devices* **2014**, *61*, 2459–2465.
5. Min, S.J.; Shin, M.C.; Nguyen, N.T.; Oh, J.M.; Koo, S.M. High-performance temperature sensors based on dual 4H-SiC JBS and SBD devices. *Materials* **2020**, *13*, 445. [CrossRef]
6. Raja, P.V.; Murty, N.V.L.N. Thermally annealed gamma irradiated Ni/4H-SiC Schottky barrier diode characteristics. *J. Semicond.* **2019**, *40*, 022804. [CrossRef]
7. Zhang, X.; Ha, S.; Benamara, M.; Skowronski, M.; O'Loughlin, M.J.; Sumakeris, J.J. Cross-sectional structure of carrot defects in 4H-SiC epilayers. *Appl. Phys. Lett.* **2004**, *85*, 5209–5211. [CrossRef]
8. Konishi, K.; Yamamoto, S.; Nakata, S.; Yu, N.; Nakanishi, Y.; Tanaka, T.; Mitani, Y.; Tomita, N.; Toyoda, Y.; Yamakawa, S. Stacking fault expansion from basal plane dislocations converted into threading edge dislocations in 4H-SiC epilayers under high current stress. *J. Appl. Phys.* **2013**, *114*, 014504. [CrossRef]
9. Maximenko, S.I.; Freitas, J.A., Jr.; Myers-Ward, R.L.; Lew, K.K.; Vanmil, B.L.; Jr, C.R.E.; Gaskill, D.K.; Muzykov, P.G.; Sudarshan, T.S. Effect of threading screw and edge dislocations on transport properties of 4H-SiC homoepitaxial layers. *J. Appl. Phys.* **2010**, *108*, 013708. [CrossRef]
10. Hayashi, S.; Yamashita, T.; Miyajima, M.; Senzaki, J.; Kato, T.; Yonezawa, Y.; Kojima, K.; Okumura, H.; Miyazato, M. Structural analysis of interfacial dislocations and expanded single Shockley-type stacking faults in forward-current degradation of 4H-SiC p-i-n diodes. *Jpn. J. Appl. Phys.* **2019**, *58*, 011005. [CrossRef]

11. Yang, Z.M.; Lan, F.; Li, Y.; Gong, M.; Huang, M.M.; Gao, B.; Hu, J.K.; Ma, Y. The effect of the interfacial states by swift heavy ion induced atomic migration in 4H-SiC Schottky barrier diodes. *Nucl. Instrum. Methods Phys. Res.* **2018**, *436*, 244–248. [CrossRef]

12. Lee, K.Y.; Huang, Y.H. An Investigation on barrier in homogeneities of 4H-SiC Schottky barrier diodes induced by surface morphology and traps. *IEEE Trans. Electron Devices* **2012**, *5*, 694–699. [CrossRef]

13. Liu, L.Y.; Shen, T.L.; Liu, A.; Zhang, T.; Bai, S.; Xu, S.R.; Jin, P.; Hao, Y.; Ouyang, X.P. Performance degradation and defect characterization of Ni/4H-SiC Schottky diode neutron detector in high fluence rate neutron irradiation. *Diam. Relat. Mater.* **2018**, *88*, 256–261. [CrossRef]

14. Huang, L.; Gu, X. Fermi level unpinning of metal/p-type 4H-SiC interface by combination of sacrificial oxidation and hydrogen plasma treatment. *J. Appl. Phys.* **2019**, *125*, 025301. [CrossRef]

15. Mandal, K.C.; Chaudhuri, S.K.; Nguyen, K.V.; Mannan, M.A. Correlation of deep levels with detector performance in 4H-SiC epitaxial Schottky barrier alpha detectors. *IEEE Trans. Nucl. Sci.* **2014**, *61*, 2338–2344. [CrossRef]

16. Yamamoto, Y.; Harada, S.; Seki, K.; Horio, A.; Mitsuhashi, T.; Koike, D.; Tagawa, M.; Ujihara, T. Low-dislocation-density 4H-SiC crystal growth utilizing dislocation conversion during solution method. *Appl. Phys. Express* **2014**, *7*, 065501. [CrossRef]

17. Yamashita, T.; Matsuhata, H.; Naijo, T.; Momose, K.; Osawa, H. Structural analysis of the 3C/4H boundaries formed on prismatic planes in 4H-SiC epitaxial films. *J. Cryst. Growth* **2016**, *455*, 172–180. [CrossRef]

18. Camarda, M.; Magna, A.L.; Via, F.L. Monte Carlo study of the early growth stages of 3C-SiC on misoriented <11-20> and <1-100> 6H-SiC substrates. *Mater. Sci. Forum* **2014**, *778*, 238–242. [CrossRef]

19. Li, Y.; Zhao, Z.; Yu, L.; Wang, Y.; Zhou, P.; Niu, Y.X.; Li, Z.H.; Chen, Y.F.; Han, P. Reduction of morphological defects in 4H-SiC epitaxial layers. *J. Cryst. Growth* **2019**, *506*, 108–113. [CrossRef]

20. Pintilie, I.; Pintilie, L.; Irmscher, K.; Thomas, B. Formation of the $Z_{1/2}$ deep-level defects in 4H-SiC epitaxial layers: Evidence for nitrogen participation. *Appl. Phys. Lett.* **2002**, *81*, 4841–4843. [CrossRef]

21. Guo, J.; Yang, Y.; Raghothamachar, B.; Kim, T.; Dudley, M.; Kim, J. Understanding the microstructures of triangular defects in 4H-SiC homoepitaxial. *J. Cryst. Growth* **2017**, *480*, 119–125. [CrossRef]

22. Lebedev, A.A. Deep level centers in silicon carbide: A review. *Semiconductors* **1999**, *33*, 107–130. [CrossRef]

23. Hu, J.; Xin, X.; Zhao, J.H.; Yan, F.; Kjornrattanawanich, B. Highly sensitive visible-blind extreme ultraviolet Ni/4H-SiC Schottky photodiodes with large detection area. *Opt. Lett.* **2006**, *31*, 1591–1593. [CrossRef]

24. Lioliou, G.; Mazzillo, M.C.; Sciuto, A.; Barnett, A.M. Electrical and ultraviolet characterization of 4H-SiC Schottky photodiodes. *Opt. Express* **2015**, *23*, 21657. [CrossRef] [PubMed]

25. Yan, G.G.; Liu, X.F.; Shen, Z.W.; Wen, Z.X.; Chen, J.; Zhao, W.S.; Wang, L.; Zhang, F.; Zhang, X.H.; Li, X.G.; et al. Improvement of fast homoepitaxial growth and defect reduction techniques of thick 4H-SiC epilayers. *J. Cryst. Growth* **2019**, *505*, 1–4. [CrossRef]

26. Lilja, L.; Hassan, J.U.; Booker, I.D.; Bergman, J.P.; Janzén, E. Influence of growth temperature on carrier lifetime in 4H-SiC epilayers. *J. Cryst. Growth* **2013**, *740–742*, 637–640. [CrossRef]

27. Storasta, L.; Tsuchida, H.; Miyazawa, T.; Ohshima, T. Enhanced annealing of the $Z_{1/2}$ defect in 4H-SiC epilayers. *J. Appl. Phys.* **2008**, *103*, 013705. [CrossRef]

28. Daigo, Y.; Ishii, S.; Kobayashi, T. Impacts of surface C/Si ratio on in-wafer uniformity and defect density of 4H-SiC homo-epitaxial films grown by high-speed wafer rotation vertical CVD. *Jpn. J. Appl. Phys.* **2019**, *58*, SBBK06. [CrossRef]

29. Watanabe, Y.; Katsuno, T.; Ishikawa, T. Relationship between characteristics of SiC-SBD and surface defect. *Hyomen Kagaku* **2014**, *35*, 84–89. [CrossRef]

30. Fujiwara, H.; Danno, K.; Kimoto, T.; Tojo, T.; Matsunami, H. Effects of C/Si ratio in fast epitaxial growth of 4H-SIC(0001) by vertical hot-wall chemical vapor deposition. *J. Appl. Phys.* **2005**, *281*, 370–376. [CrossRef]

31. Miyazawa, T.; Tsuchida, A.H. Growth of 4H-SiC epilayers and $Z_{1/2}$ center elimination. *Mater. Sci. Forum* **2012**, *717–720*, 81–86. [CrossRef]

32. Yazdanfar, M.; Stenberg, P.; Booker, I.D.; Ivanov, I.G.; Kordina, O.; Pedersen, H.; Janzen, E. Process stability and morphology optimization of very thick 4H-SiC epitaxial layers grown by chloride-based CVD. *J. Cryst. Growth* **2013**, *113*, 125–130. [CrossRef]

33. Shrivastava, A.; Muzykov, P.; Caldwell, J.D.; Sudarshan, T.S. Study of triangular defects and inverted pyramids in 4H-SiC 4° off-cut (0001) Si face epilayers. *J. Cryst. Growth* **2008**, *310*, 4443. [CrossRef]

34. Feng, G.; Suda, J.; Kimoto, T. Characterization of major in-grown stacking faults in 4H-SiC epilayers. *Phys. Lett.* **2009**, *94*, 091910. [CrossRef]

35. Niwa, H.; Feng, G.; Suda, J.; Kimoto, T. Breakdown characteristics of 15-kV-class 4H-SiC PiN diodes with various junction termination structures. *IEEE Trans. Electron Devices* **2012**, *59*, 2748–2752. [CrossRef]

36. Xin, B.; Jia, R.X.; Hu, J.C.; Tsai, C.Y.; Lin, H.H.; Zhang, Y.M. A step-by-step experiment of 3C-SiC hetero-epitaxial growth on 4H-SiC by CVD. *Appl. Surf. Sci.* **2015**, *357*, 985–993. [CrossRef]

37. Chen, W.; Lee, K.; Capano, M. Growth and characterization of nitrogen-doped C-face 4H-SiC epilayers. *J. Cryst. Growth* **2006**, *297*, 265–271. [CrossRef]

38. Kojima, K.; Okumura, H.; Kuroda, S.; Arai, K. Homoepitaxial growth of 4H-SiC on -axis (0001(_)) C-face substrates by chemical vapor depositon. *J. Cryst. Growth* **2004**, *269*, 367–376. [CrossRef]

39. Leone, S.; Beyer, F.C.; Pedersen, H.; Kordina, O.; Henry, A.; Janzen, E. Growth of smooth 4H-SiC epilayers on $4°$ off-axis substrates with chloride-based CVD at very high growth rate. *Mater. Res. Bull.* **2011**, *46*, 1272–1275. [CrossRef]

40. Pristavu, G.; Brezeanu, G.; Badila, M.; Pascu, R.; Danila, M.; Godignon, P. A model to non-uniform Ni Schottky contact on SiC annealed at elevated temperatures. *Appl. Phys. Lett.* **2015**, *106*, 261605. [CrossRef]

41. Kyoung, S.; Jung, E.S.; Kang, T.Y.; Sung, M.Y. Optimized designing to improve electrical characteristics of 4H-SiC wide trench junction barrier Schottky diode. *Sci. Adv. Mater.* **2018**, *10*, 416–421. [CrossRef]

42. Okino, H.; Kameshiro, N.; Konishi, K.; Shima, A.; Yamada, R. Analysis of high reverse currents of 4H-SiC Schottky-barrier diodes. *J. Appl. Phys.* **2017**, *122*, 235704. [CrossRef]

43. Benamara, M.; Anani, M.; Akkal, B.; Benamara, Z. Ni/SiC-6H Schottky barrier diode interfacial states characterization related to temperature. *J. Alloys Compd.* **2014**, *603*, 197–201. [CrossRef]

44. Kzlovski, V.V.; Lebedev, A.A.; Levinshtein, M.E.; Rumyantsev, S.L.; Palmour, J.W. Impact of high energy electron irradiation on high voltage Ni/4H-SiC Schottky diodes. *Appl. Phys. Lett.* **2017**, *110*, 133501. [CrossRef]

45. Shabunina, E.I.; Levinshtein, M.E.; Shmidt, N.M.; Ivanov, P.A.; Palmour, J.W. 1/f noise in forward biased high voltage 4H-SiC Schottky diodes. *Solid State Electron.* **2014**, *96*, 44–49. [CrossRef]

46. Zhang, C.X.; Zhang, E.X.; Fleetwood, D.M.; Schrimpf, R.D.; Dhar, S.; Ryu, S.H.; Shen, X.; Pantelides, S.T. Origins of low-frequency noise and interface traps in 4H-SiC MOSFETs. *IEEE Electron Device Lett.* **2013**, *34*, 117–119. [CrossRef]

47. Alexander, S.; David, Z.Y.; Ting, C.J.; Hill, M.L.; Jean, N.; Sam, A.; Keo, J.M.; Mumolo and Sarath, D.G. Gain and noise of high-performance long wavelength superlattice infrared detectors. *Appl. Phys. Lett.* **2010**, *96*, 1–3.

48. Kalinina, E.V.; Violina, G.N.; Nikitina, I.P.; Yagovkina, M.A.; Zabrodski, V.V. Proton irradiation of 4H-SiC photodetectors with Schottky barriers. *Semiconductors* **2019**, *53*, 844–849. [CrossRef]

49. Litton, C.W.; Johnstone, D.; Akarca-Biyikli, S.; Ramaiah, K.S.; Bhat, I.; Chow, T.P.; Kim, J.K.; Schubert, E.F. Effect of C/Si ratio on deep levels in epitaxial 4H-SiC. *Appl. Phys. Lett.* **2006**, *88*, 121914. [CrossRef]

50. Eberlein, T.A.G.; Jones, R.; Briddon, P.R. Z_1/Z_2 defects in 4H-SiC. *Phys. Rev. Lett.* **2003**, *90*, 225502. [CrossRef]

51. Trinh, X.T.; Szasz, K.; Homos, T.; Kawahara, K.; Suda, J.; Kimoto, T.; Gali, A.; Janzén, E.; Son, N.T. Negative-U carbon vacancy in 4H-SiC: Assessment of charge correction schemes and identification of the negative carbon vacancy at the quasicubic site. *Phys. Rev. Lett.* **2013**, *88*, 235209. [CrossRef]

52. Lilja, L.; Booker, I.D.; Hassan, J.U.; Janzen, E.; Bergman, J.P. The influence of growth conditions on carrier lifetime in 4H-SiC epilayers. *J. Cryst. Growth* **2013**, *381*, 43–50. [CrossRef]

 *micromachines*

Article

A Graphene/Polycrystalline Silicon Photodiode and Its Integration in a Photodiode–Oxide–Semiconductor Field Effect Transistor

Yu-Yang Tsai, Chun-Yu Kuo, Bo-Chang Li, Po-Wen Chiu and Klaus Y. J. Hsu *

Institute of Electronics Engineering, National Tsing Hua University, Hsinchu 30013, Taiwan;
st95410@gmail.com (Y.-Y.T.); kuochenyou@gmail.com (C.-Y.K.); pet791228@gmail.com (B.-C.L.);
pwchiu@ee.nthu.edu.tw (P.-W.C.)
* Correspondence: yjhsu@ee.nthu.edu.tw; Tel.: +886-3-574-2594

Received: 25 May 2020; Accepted: 15 June 2020; Published: 17 June 2020

Abstract: In recent years, the characteristics of the graphene/crystalline silicon junction have been frequently discussed in the literature, but study of the graphene/polycrystalline silicon junction and its potential applications is hardly found. The present work reports the observation of the electrical and optoelectronic characteristics of a graphene/polycrystalline silicon junction and explores one possible usage of the junction. The current–voltage curve of the junction was measured to show the typical exponential behavior that can be seen in a forward biased diode, and the photovoltage of the junction showed a logarithmic dependence on light intensity. A new phototransistor named the "photodiode–oxide–semiconductor field effect transistor (PDOSFET)" was further proposed and verified in this work. In the PDOSFET, a graphene/polycrystalline silicon photodiode was directly merged on top of the gate oxide of a conventional metal–oxide–semiconductor field effect transistor (MOSFET). The magnitude of the channel current of this phototransistor showed a logarithmic dependence on the illumination level. It is shown in this work that the PDOSFET facilitates a better pixel design in a complementary metal–oxide–semiconductor (CMOS) image sensor, especially beneficial for high dynamic range (HDR) image detection.

Keywords: graphene; polycrystalline silicon; photodiode; phototransistor; pixel; high dynamic range (HDR) image

---

1. Introduction

The unique structural, electronic, and optical properties of graphene have made this two-dimensional material an attractive subject of research in recent years. While the high carrier mobility in graphene opened the door to its applications in high-speed electronics [1–4], the application of graphene in optoelectronics has also become an interesting and important topic. The linear dispersion of the Dirac electrons in graphene can be utilized for broadband optical detection, and the intrinsic speed of the photodetectors can be very high. Unfortunately, using single-layer graphene for vertical optical absorption led to very low photocurrent responsivities [5–8], normally below or about 10 mA/W, because of the high transparency of graphene over a wide band of light wavelengths [9,10]. Many interesting strategies have been proposed to enhance the optical absorption by the graphene in graphene-based photodetectors. They require specially patterned structures such as waveguides [11], resonant cavities [12], plasmonic nanostructures [13], graphene nanodisks [14], or graphene quantum dots [15,16], and the resultant photocurrent responsivities have been successfully increased. For more responsive optical detection in simple planar, single-layer graphene-based photodetectors, a natural way is to construct a hybrid device in which the role of optical absorber is left to other materials in the device and the graphene is used as an efficient conductive material for carrier collection. Graphene has been adopted to form transparent electrodes in various solar cells [17–20]. It has also been extensively combined with crystalline silicon (c-Si) and germanium to form Schottky photodiodes [21–32]. In these cases, the transparency of graphene becomes an advantage. Incident light is not blocked by the graphene, and the spectral photocurrent responsivities of the devices are determined by the other combined materials instead of the graphene. Taking graphene/c-Si junction photodetectors as an example, the c-Si is the optical absorber and photocurrent responsivities in the order of 100 mA/W are typically obtained. The performance is similar to that of typical c-Si P/N junction photodiodes and is good enough for numerous applications.

Interestingly, while there have been a number of studies studying graphene/c-Si junctions and their application to photodetection [21–31], the study of the graphene/polycrystalline silicon (graphene/poly-Si) junction and its application is rare in the literature. Only Lin et al. have investigated the change in the current density-voltage (J-V) characteristic of a four-layer graphene/p-type poly-Si junction after being influenced by some ultraviolet illumination [33]. No study on single-layer graphene/n-type poly-Si junctions has been reported. However, it may be beneficial to pay attention to this kind of junction because it could have application potential on occasions when poly-Si exists. Metal–oxide–semiconductor field effect transistors (MOSFETs) with a poly-Si gate and poly-Si thin film solar cells are two examples of possible applications. In the present work, a junction made of single-layer graphene and n-type poly-Si was fabricated and characterized. Furthermore, a new phototransistor named the PD–oxide–semiconductor field effect transistor (PDOSFET), in which the graphene/poly-Si junction photodiode (PD) is embedded in a conventional metal–oxide–semiconductor field effect transistor (MOSFET), was proposed and experimentally verified. Analysis showed that such phototransistor is particularly useful in constructing the pixels in complementary-metal–oxide–semiconductor (CMOS) image sensors for detecting high-contrast images in which the brightness is of high dynamic range (HDR).

2. Fabrication and Characterization of Graphene/Poly-Si Junction

Before a graphene/poly-Si junction is used for applications, it is essential to have a look at its general electrical and optoelectronic characteristics. For this purpose, samples of graphene/n-type poly-Si junction were fabricated. A 300 nm-thick phosphorus-doped poly-Si film was deposited on an n-type Si wafer by using low pressure chemical vapor deposition (LPCVD) at a temperature of 620 °C. Because the optical absorption coefficient of poly-Si in the visible light band is between 10^4 and 10^5 cm^{-1} [34], the thickness (300 nm) of the poly-Si was chosen to ensure that most of the incident photons could be absorbed when the film was illuminated by visible light. The graphene film was first grown on a copper foil by using ambient pressure chemical ovapor deposition (APCVD) at a temperature of 1040 °C and was subsequently transferred onto the poly-Si by using a commonly adopted graphene wet transfer process [35,36]. The poly-Si was partially covered by the transferred graphene film. Afterwards, the wafer was cut into discrete devices, each with an area of about 25 mm^2. Silver paste was respectively applied to the surface of exposed poly-Si and the surface of graphene to form the two electrical terminals of the graphene/n-type poly-Si junction. The reason for using silver paste as the contact material was that it could be easily applied to quickly finish sample preparation and it had been tested to form ohmic contacts with graphene and with n-type c-Si. It was speculated that the paste might still form ohmic contacts with n-type poly-Si. Figure 1 shows a photo of typical samples placed on glass.

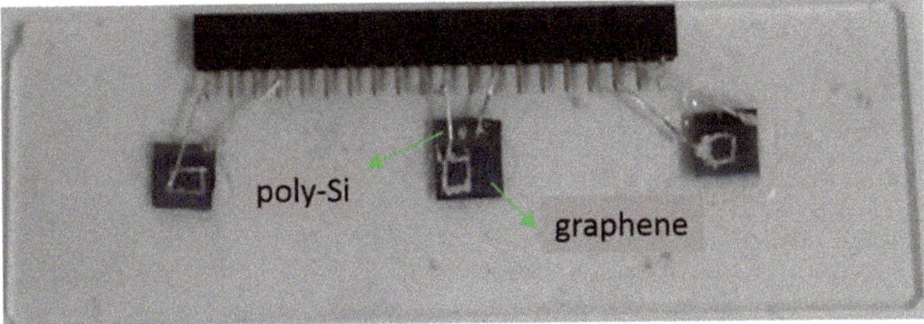

Figure 1. Samples of graphene/poly-Si junction. The graphene film partially covers the poly-Si in each sample.

The sheet resistance (R_{SH}) of the graphene film was measured by using a four-point probe. Values between 250 Ω/square and 300 Ω/square were obtained for all the samples. Figure 2 shows the Raman spectrum of the graphene after the transfer process. From the 2:1 intensity ratio between the 2D and G peaks, it can be deduced that the transferred graphene is of a single layer [35]. The slight D peak indicates that there may be some defects such as wrinkles in the transferred graphene layer.

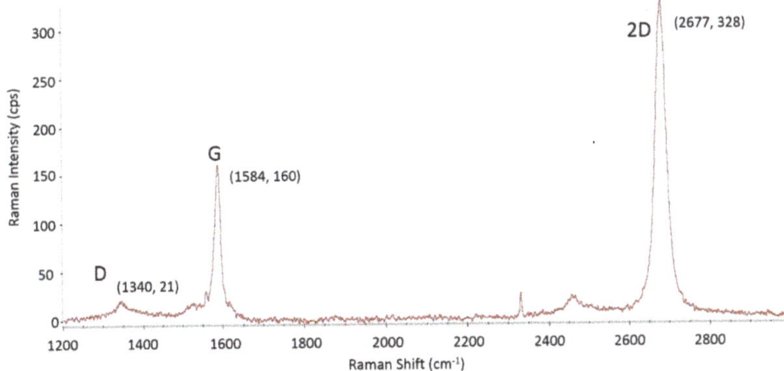

Figure 2. Raman spectrum of the transferred graphene layer. It indicates that the graphene is a single layer with slight defects.

Optical measurement was also performed to characterize the optical transmission rates of several transferred graphene samples. Graphene films were placed on glass substrates with a 96% transmission rate. The total transmission rates of the samples were then measured. All the resultant transmission rates of the graphene films are between 97% and 98% over a wide range of optical wavelengths, which is consistent with the single-layer structure of the graphene films. Figure 3 shows a typical measured spectrum of the graphene-on-glass assembly.

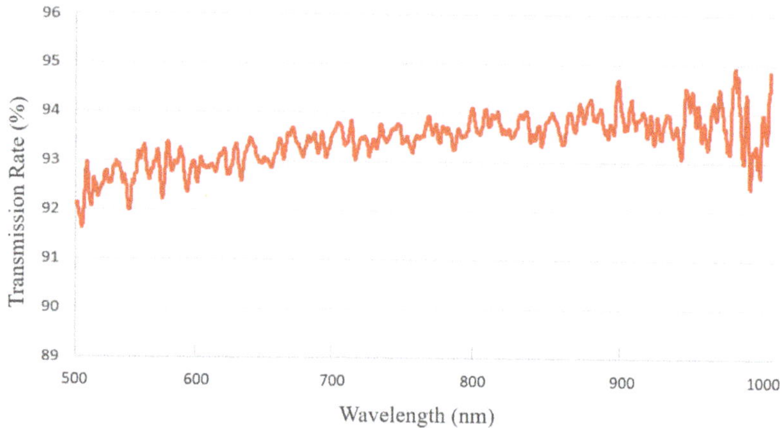

Figure 3. Measured total transmission rate of a typical graphene-on-glass sample. The transmission rate of the glass is about 96%.

To see the graphene/n-type poly-Si junction behave like an ohmic junction or a Schottky junction, the dark *I–V* curve of the fabricated graphene/n-type poly-Si junction was measured by using the Agilent 4156a Semiconductor Parameter Analyzer (Agilent, Santa Clara, CA, USA) at room temperature with the terminal of the poly-Si set at ground potential. Figure 4 shows a typical result along with the *I–V* curve under the illumination of a halogen lamp. The optical power of the lamp was unknown. It can be seen from Figure 4 that the measured *I–V* curves do not show the simple exponential behavior of a single rectifying junction. Instead, they show the typical behavior of a metal–semiconductor–metal (MSM) structure with two asymmetric rectifying junctions [37,38]. The exponential behavior of the curves under negative bias indicates that in addition to the single-layer graphene/n-type poly-Si junction, another rectifying junction with a smaller built-in potential and in the opposite direction exists in the current path. In this case, it should be due to the silver paste/poly-Si junction. Unlike our previously tested silver paste/n-type c-Si junction that showed ohmic conduction, this silver paste/poly-Si junction appears to be a Schottky contact, and it forms an MSM structure along with the single-layer graphene/n-type poly-Si junction. The phenomenon that the current under forward bias increased when light was applied to the sample also confirms this point. The reverse biased silver paste/poly-Si junction contributed a photo-generated electron current to the forward biased graphene/n-type poly-Si junction. If the silver paste/poly-Si junction was ohmic, the forward bias current should have decreased when the light was on. Although the silver paste/poly-Si junction complicates the measured *I–V* curves and it is no longer straightforward to extract the exact values of junction parameters (e.g., Richardson constant and Schottky barrier), we can still see that the dark current flowing through the single-layer graphene/n-type poly-Si junction increases exponentially with increasing forward bias voltage. The large ideality factor, *n*, of this disturbed forward bias characteristic was found by curve fitting to be about 10. The exponential dark *I–V* curve under forward bias implies that we should be able to see a logarithmic dependence on light intensity from the open-circuit photovoltage of the junction, as analyzed below.

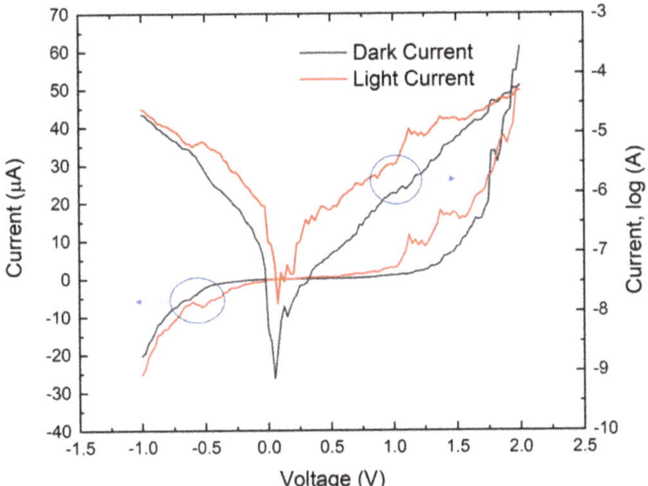

Figure 4. Measured *I–V* curves of the graphene/n-type poly-Si junction. The voltage is applied to the graphene terminal and the poly-Si terminal is grounded.

The general exponential dark *I–V* characteristic of a junction can be expressed as

$$I_D = I_S\left[\exp\left(\frac{V_D}{V_T}\right) - 1\right] \tag{1}$$

where I_D is the junction current, $V_T = \frac{nkT}{q}$, n is the ideality factor of the junction, I_s is the dark reverse saturation current of the junction, and V_D is the voltage across the junction. When this junction is used as a PD, the short-circuit photocurrent (I_{ph}) of the PD is directly proportional to the incident light intensity. If at $t = 0$, a light starts to illuminate the junction under the open-circuit condition, the photocurrent does not flow out. It charges the junction capacitance (C_D) and internally forward biases the junction at the same time. The dynamics can be described by

$$I_{ph} - I_D = C_D\frac{dV_D}{dt} \tag{2}$$

Solving Equations (1) and (2) by the separation of the variables, with the zero initial condition $V_D(0) = 0$, one can derive that the open-circuit junction voltage under illumination, i.e., the photovoltage (V_{ph}), should follow the next equation:

$$V_{ph}(t) = V_T \ln\left\{\frac{I_{ph} + I_s}{I_{ph}\exp[-\frac{(I_{ph}+I_s)t}{V_T C_D}] + I_s}\right\} \tag{3}$$

where I_{ph} is the photocurrent, and t is the exposure time. The logarithmic response of the photovoltage to light intensity (via I_{ph}) is advantageous because it mimics human eyes' response to brightness. Utilizing V_{ph} as the output signal allows one to detect a wide dynamic range of light intensity. Unlike the photocurrent that may easily saturate the subsequent reading circuit at high illuminance, the photovoltage allows the circuit to function properly because the signal is logarithmically compressed at high illuminance but still linear at low illuminance. When the incident light is weak or the exposure time is short, Equation (3) reduces to

$$V_{ph} \cong \frac{I_{ph}t}{C_D} \tag{4}$$

which indicates that the photovoltage under the weak illumination condition is a linear function of the incident light intensity, just as the photocurrent is. On the other hand, for strong illumination that produces a large photocurrent, Equation (3) becomes

$$V_{ph} = V_T \ln\left(\frac{I_{ph} + I_s}{I_s}\right) \cong V_T \ln\left(\frac{I_{ph}}{I_s}\right) \tag{5}$$

which shows that the photovoltage is a compressed output signal of the strong incident light intensity. Note that both Equations (4) and (5) are independent of the area of PD since I_{ph}, I_s, and C_D are all proportional to the junction area, which means that it is not necessary to enlarge the device size of the PD in order to obtain a significant magnitude of photovoltage.

Since the measured dark I–V curve of the fabricated graphene/n-type poly-Si junction sample at room temperature shows an exponential characteristic under forward bias, it can be expected that if this junction is used as a PD, the photovoltage of the PD should exhibit a logarithmic dependence on the incident light intensity. Even though there exists an additional silver paste/poly-Si junction at the other side of the poly-Si, the photo-generated holes in the poly-Si can still flow to the graphene and build up the photovoltage. Figure 5 shows the measured open-circuit photovoltage of a graphene/n-type poly-Si junction sample under the illumination of a halogen lamp. When the illuminance is 49,000 lux, the photovoltage reaches 0.424 V. As expected, the photovoltage increases logarithmically with the illuminance. At higher illuminance, the increase in photovoltage is gradually compressed.

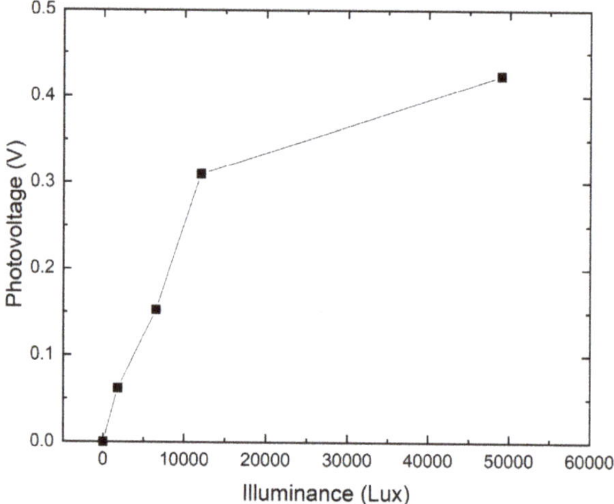

Figure 5. Measured photovoltage of the graphene/n-type poly-Si junction photodiode (PD) under the illumination of a halogen lamp.

3. Photodiode–Oxide–Semiconductor Field Effect Transistor

The logarithmic behavior of the photovoltage of a PD makes this variable particularly useful for being adopted in CMOS image sensors as a signal to detect high dynamic range (HDR) images. Furthermore, the graphene/n-type poly-Si junction PD can bring even more benefit to this application because it can be implemented right on top of the gate oxide of a MOSFET in the pixel of a CMOS image sensor. Figure 6 shows the typical structure of a three-transistor (3T) active pixel commonly used in a CMOS image sensor [39–41].

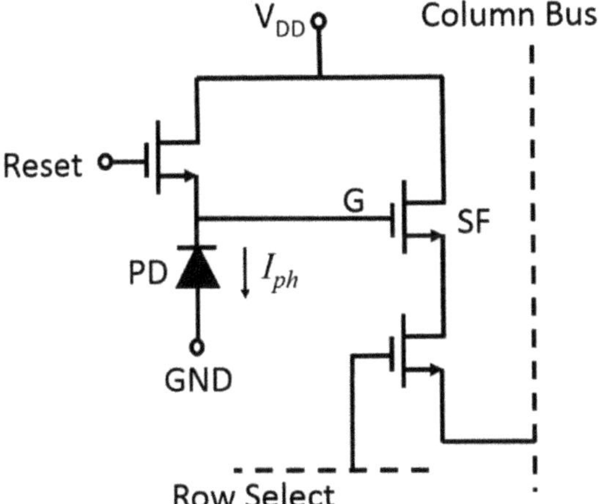

Figure 6. Conventional 3T pixel structure in a complementary-metal–oxide–semiconductor (CMOS) image sensor.

Conventionally, the PD in the pixel is a monolithic P–N junction diode located beside the transistors. To operate the pixel, the Reset transistor is first turned on to set the voltage at node G (V_G) to a high value (V_{DD}). Then, the Reset transistor is turned off and V_G is reduced by the discharging mechanism due to the photocurrent I_{ph} generated from the PD. The amount of the voltage reduction (ΔV) at node G is a linear function of I_{ph} as shown in Equation (6):

$$\Delta V = \frac{I_{ph}\Delta t}{C_T} = \frac{I_{ph}\Delta t}{C_D + C_G + C_p} \tag{6}$$

where Δt is a specified discharging period, C_T is the total capacitance at node G, C_D is the capacitance of the PD, C_G is the capacitance looking into the gate of the source follower (SF) transistor, and C_p is the parasitic capacitance due to interconnect. A large fill factor of the PD in the pixel is preferred so that I_{ph} can be maximized and C_T can be less influenced by C_G and C_p. After the discharging period,

$$V_G = V_{DD} - \Delta V \tag{7}$$

The V_G signal is output through the source follower to the column bus, and ΔV is used to represent the average light intensity during the discharging period. This linear operation principle has limitations: Firstly, for high light intensity, I_{ph} and ΔV can easily become too large and V_G may be reduced too much to make the SF transistor functional. This saturation limits the detectable range of input light intensity. Secondly, for high resolution image sensors, there exists a tradeoff between the pixel size and the fill factor of the PD in a pixel. This limits the values of I_{ph} and C_D, and the influence from C_G and C_p becomes obvious. Modifying the 3T pixel structure can resolve these limitations.

Figure 7 shows the proposed modified 3T pixel structure in which only slight changes from Figure 6 are needed: the location of the Reset switch is changed, the polarity of the PD is reversed, and the ground (GND) potential is replaced by a proper positive bias voltage (V_{REF}) that makes the SF transistor always functional. In this way, the V_G is first reset to V_{REF} and then becomes the sum of V_{REF} and the photovoltage of the PD after some exposure time t. The photovoltage of the PD can be expressed as Equation (3) except that C_D should now be replaced by C_T. Therefore, as explained in Equations (4) and (5), the V_G in the modified pixel structure now behaves linearly at low light intensity but changes logarithmically at high light intensity. With the proper setting of V_{REF}, the V_G value does not exceed V_{DD} and can always make the SF transistor functional. This natural compression of the signal allows the dynamic range of incident light intensity to be extremely wide, which makes the single-shot detection of high-contrast images possible. There is no need to take multiple shots of an HDR image at different exposure levels and then blend these images together for a composite image. Furthermore, if the original P–N junction PD is replaced with a graphene/n-type poly-Si junction PD that is directly fabricated on top of the gate oxide of the SF transistor, the parasitic capacitance C_p can be eliminated and the tradeoff between pixel size and fill factor can be removed. That is, one can combine the PD and the SF transistor into a new integrated device, namely, the PD–oxide–semiconductor field effect transistor (PDOSFET), to maximize pixel performance. In this way, since the PD is directly stacked on the gate of the SF transistor, metallic interconnect is not needed and C_p is completely eliminated. In addition, because the stacking makes the area of the PD the same as the gate area of the SF transistor, the photovoltage of the PD is independent of the area and there is no more area-competing issue between the PD and the MOSFET. Poly-Si/graphene PD offers unique advantages for this integration. The n-type poly-Si gate is an essential element in most CMOS technologies. All one needs to do is to apply a graphene film to the gate area to form the PD. Additionally, the graphene film has good thermal stability and can withstand the thermal cycles in subsequent fabrication processes. Therefore, one can apply it either before or after the poly-Si deposition process, which allows flexibility in designing the polarity of the PD. If graphene is placed above poly-Si, its transparency allows light to pass through and reach the poly-Si for optical absorption. Metal/poly-Si Schottky PD cannot provide these advantages.

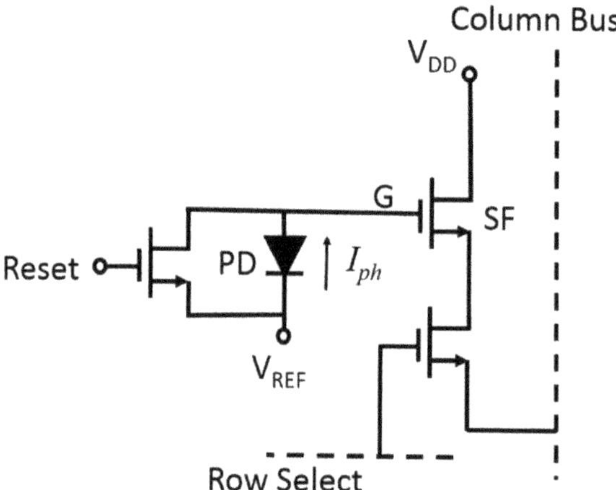

Figure 7. Proposed 3T pixel structure in a CMOS image sensor.

To verify the PDOSFET concept, a test device as schematically illustrated in Figure 8 was fabricated. The channel length was set to be long (6 µm), and the channel width was designed to be 12 µm. The gate oxide was grown by thermal oxidation after the formation of a P-well. It was chosen to be thick (20 nm) to reduce leakage currents in order to assure that the photovoltage developed by the PD would not be degraded. A single-layer graphene was transferred onto the gate oxide before depositing the 300 nm-thick n-type poly-Si gate.

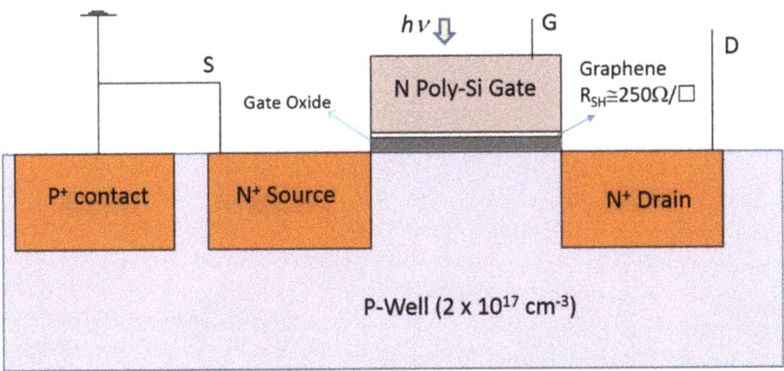

Figure 8. Schematic illustration of a graphene photodiode–oxide–semiconductor field effect transistor (PDOSFET) test device. Poly-Si gate thickness: 300 nm; gate oxide thickness: 20 nm; gate length: 6 µm.

Figure 9 shows the process of transferring the single-layer graphene from copper foil onto the gate oxide. After the graphene was synthesized on the copper foil, it was first supported and protected by a coated poly(bisphenol A carbonate) (PC) layer, and the redundant graphene on the back side of the copper foil was etched away by using reactive ion etching (RIE) with oxygen plasma. Then, the copper foil was removed in an aqueous solution of HCl and H_2O_2. After rinsing in de-ionized (DI) water, the graphene was placed onto the Si wafer with gate oxide already grown on it, and the PC layer was subsequently removed with chloroform. The graphene was then patterned by lithography and RIE according to the designed gate pattern, followed by normal poly-Si gate deposition, lithography, and self-aligned CMOS processes in which the source and drain doping was implemented by using arsenic (As) ion implantation with an energy of 30 KeV and a dose of 8×10^{13} cm⁻². Because front illumination is adopted for the PDOSFET to detect light, the gate contact metal was intentionally designed to cover only a very small portion of the gate area so that light could reach the graphene/n-type poly-Si junction PD. However, both the drain and source regions were fully covered by a metal layer so that their parasitic P–N junction diodes would not intervene during illumination. The ohmic contact to the poly-Si gate was formed by sequentially depositing Ti and Au films. Therefore, the MSM structure observed in the samples shown in Section 2 did not appear in the PDOSFET. The graphene film sustained the subsequent processes. The poly-Si of one sample was removed to expose the underlying graphene for Raman spectroscopic characterization. The same spectrum as that in Figure 2 was observed. The single-layer graphene held its structure well.

The resultant PDOSFET looks just like a normal n-channel MOSFET (NMOSFET) except that a graphene layer is inserted between the n-type poly-Si gate and the gate oxide. That is, a PD is embedded right in the gate structure with minimal process modification. The device should be able to exhibit the normal dark electrical characteristics of an NMOSFET. To verify this, the I_d-V_g and I_d-V_d characteristics of the fabricated PDOSFET in the dark were measured by using an Agilent 4156a semiconductor parameter analyzer. Figure 10 shows the measured I_d–V_g curve of the PDOSFET under 3.3 V drain-to-source voltage (V_{ds}). Normal turn-on behavior was observed. The threshold voltage for the channel to turn on is between 1 V and 1.5 V. By using the ATLAS device simulator from Silvaco, the threshold voltage of a MOSFET, the same as that for the PDOSFET shown in Figure 8 but without the embedded graphene layer, was calculated to be about 0.7 V. This MOSFET control device was fabricated and measured, and it showed a threshold voltage of about 0.85 V. The larger threshold voltage of the PDOSFET may be due to the existence of a depletion region in the poly-Si side of the graphene/poly-Si diode. It may absorb part of the applied gate voltage (V_g) so that the control of the surface potential at the channel surface by V_g becomes less effective. Figure 11 shows the measured I_d–V_d characteristics of the PDOSFET under various gate bias voltages. Normal linear and saturation behaviors can be clearly recognized. The channel length of the fabricated PDOSFET is quite long; therefore, the short channel effect is hardly seen in Figure 11. From Figures 10 and 11, it can be confirmed that the insertion of a graphene layer into the gate of a MOSFET does not destroy the properties of the transistor.

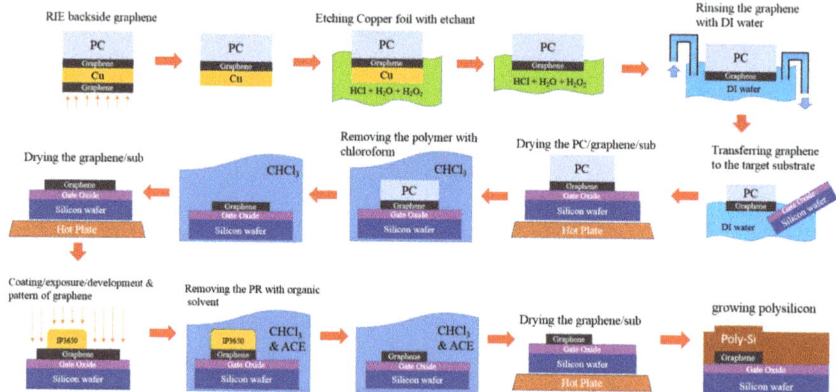

Figure 9. Schematic illustration of the graphene transfer process in fabricating a PDOSFET.

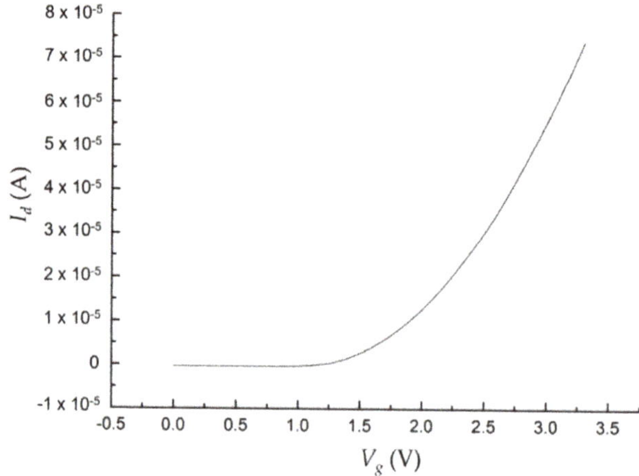

Figure 10. Measured dark turn-on characteristic (I_d–V_g curve) of the PDOSFET; V_{ds} = 3.3 V.

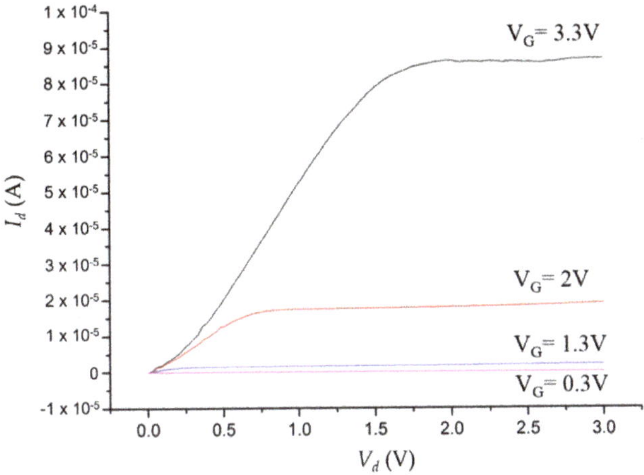

Figure 11. Measured dark I_d–V_d curve of the PDOSFET under various gate bias voltages.

When the PDOSFET is illuminated with light, the photovoltage developed at the PD should be able to effectively increase the gate voltage of the FET and thus increase the channel current. Therefore, when the PDOSFET is connected in a source follower configuration, the output voltage of the source follower should faithfully reflect the logarithmic behavior of the photovoltage of the PD. Figure 12 shows the measurement result for such a source follower when the PDOSFET was illuminated by a halogen lamp. The source follower connection is illustrated in the inset of Figure 12. It is clearly seen that the source follower output voltage (V_{out}) shows a logarithmic dependence on illuminance. This indicates that both the graphene/n-type poly-Si junction PD and the FET integrated in the PDOSFET functioned well. Under low illuminance, V_{out} changes linearly with light intensity to provide good resolution. At high illuminance, V_{out} is compressed to prevent saturation. The feasibility of using a single PDOSFET in a pixel to replace the P–N junction PD and the SF transistor is thus verified. It should be mentioned that the modification from the conventional 3T pixel structure to the new 3T pixel structure with the PDOSFET is very slight and that the responsivity of the graphene/poly-Si PD is mainly determined by poly-Si, not by graphene. Therefore, except for the optical input dynamic range, most other performance parameters such as the quantum efficiency, conversion gain, noise, and image lag of the new pixel should be similar to those of the conventional pixel.

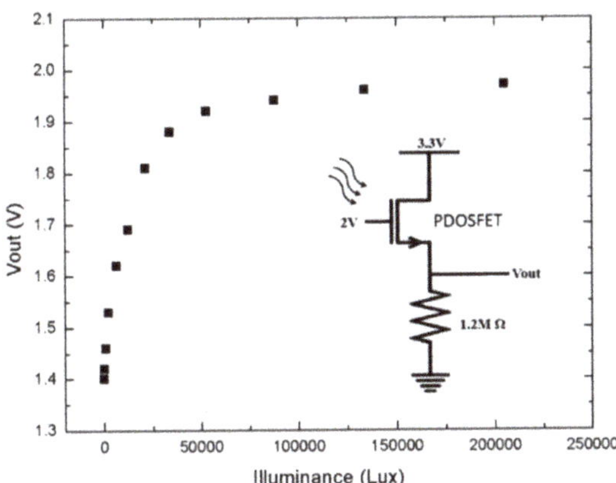

Figure 12. Measured PDOSFET source follower output voltage under various illuminance values.

Although the thickness of the poly-Si gate was designed to be 300 nm to absorb most of the visible light, it cannot be ruled out that some long wavelength light may still travel through the gate and reach the c-Si region under the gate oxide. These photons may still generate electron–hole pairs and contribute to the increase in the drain current. However, if this influence exists and dominates, the magnitude of the drain current increase should be linearly proportional to the increasing illuminance. Besides, the photo-generated drain current of the MOSFET control device (without PD at the gate) under 49,000 Lux illumination was measured to be only 0.1 µA even at V_{DS} = 5 V. Therefore, the observed logarithmic dependence of V_{out} (thus, the channel current) on illuminance is certainly not due to (at least not dominated by) the optical absorption in the c-Si substrate. Instead, the graphene/n-type poly-Si PD on top of the gate dominates the optoelectronic behavior of the PDOSFET source follower.

In addition, it was mentioned above that the source and drain areas of the PDOSFET were covered by metal in order to prevent the source and drain junctions from detecting light and contributing extra photocurrent. The effectiveness of this arrangement was investigated by measuring the dark and light drain currents when the gate voltage was set low to disable the channel. No change in the terminal current magnitude was observed with and without illumination. This confirms that the metal coverage on the source and drain regions effectively blocked the incident light.

For most applications, it is desirable for the intrinsic operation speed of the phototransistor to be fast. Therefore, the transient response of the PDOSFET source follower was also investigated in this work. A white light-emitting diode (LED) was switched on/off by a voltage square wave to produce 5000 Lux light pulses onto the PDOSFET. The transient behavior of the source follower output voltage (V_{out}) was observed. For the fabricated large, long-channel PDOSFET, both the rise time and the fall time of V_{out} were as short as 22 µs when the loading resistance was 100 kΩ, as shown in Figure 13. When the loading resistance was increased to 1.2 MΩ, the rise time and the fall time only increased to 56 µs and 174 µs, respectively. This fast speed is beneficial to many applications.

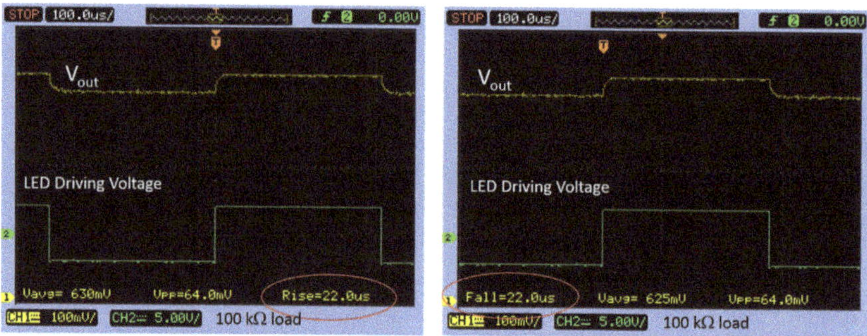

Figure 13. Measured transient behavior of the output voltage (V_{out}) of the PDOSFET source follower with a loading resistance of 100 kΩ.

One process issue about employing the proposed PDOSFET in mass production is the use of graphene in the front-end of standard, commercial CMOS processes. In such mass production processes, it is impossible to deposit the graphene layer by using slow techniques such as the transfer method. The direct wafer-level synthesis/deposition of graphene is needed. Besides, possible carbon contamination of the devices on the chip, due to the gases used in the graphene deposition process, must be avoided. Otherwise, graphene will not be usable in the front end of CMOS processes and it will be impossible to fabricate the PDOSFET. Although the new pixel structure shown in Figure 7 can still be implemented by using separate P–N junction diode and source follower MOSFET instead of using the PDOSFET, it is certain that the signal amplitude will be smaller and that the operation speed of the pixel will be slower because more parasitic capacitance emerges, and the conventional area competing problem between the PD and transistors will remain. Thus, the PDOSFET holds an important position in efficient HDR image detection. It is best for relevant front-end processes to be developed as soon as possible to facilitate the deployment of the PDOSFET.

As a remark, one possible way to implement the pixel structure shown in Figure 7 without using the PDOSFET to resolve the fill factor issue of the pixel is to use graphene/amorphous-silicon (Gr/a-Si) junctions as the PDs and fabricate these graphene/amorphous-silicon junctions during or after the back-end processes. In this way, although graphene is used, one can avoid producing contamination in front-end processes. The low-temperature process of depositing a-Si films also has minimal impact on the underlying CMOS devices. Since the graphene/amorphous-silicon PD and the transistors in the pixel do not locate in the same plane, they do not need to compete for chip area anymore. To demonstrate this alternative idea, we have designed and fabricated a Gr/n-type a-Si PD located above a Si NMOSFET, as shown in Figure 14. The graphene terminal of the Gr/n-type a-Si photodiode was connected to the poly-Si gate of the NMOSFET through a metal via. There is much more freedom for designing the area of the PD. In our samples, the a-Si was first deposited by using a hot wire chemical vapor deposition (HWCVD) system without doping. Its film thickness was about 150 nm. Then, phosphorus ion implantation was performed at 110 KeV of energy and a 1×10^{12} cm^{-2} dose to make it n-type. At the ohmic contact area, heavy doping was achieved by an additional ion implantation with 10 KeV of energy and a 2×10^{15} cm^{-2} dose. The area of the PD was designed to be 29.5×66 μm^2, much larger than the gate area (6×12 μm^2) of the NMOSFET, to enhance the coupling of photovoltage to the gate of the NMOSFET. Figure 15 shows that the source follower circuit based on the sample in Figure 14 functioned very well. The logarithmic characteristic of the photovoltage was successfully revealed at the output node. Compared with the source follower circuit based on the PDOSFET as shown in Figure 12, the range of the output voltage span of this source follower circuit is smaller. This is not surprising, since metallic interconnect introduced more parasitic capacitance, which reduced the photovoltage magnitude. It may be possible to compensate this photovoltage reduction by enlarging the PD area so that its junction capacitance can dominate.

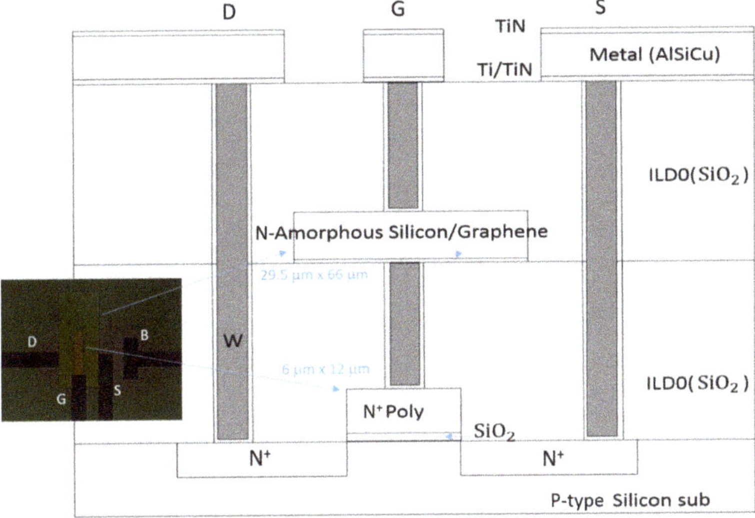

Figure 14. Schematic illustration of the Gr/a-Si PD and n-channel MOSFET (NMOSFET) connected structure. The PD is located above the NMOSFET, and the area of the PD is larger than that of the NMOSFET.

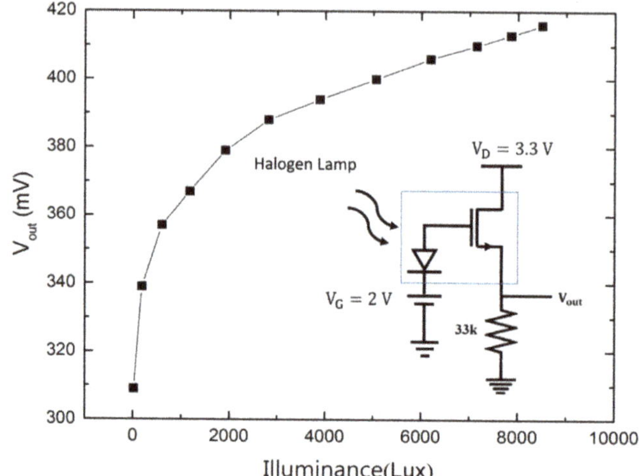

Figure 15. Measured Gr/a-Si PD and NMOSFET source follower output voltage under various illuminance.

4. Conclusions

A graphene/n-type poly-Si junction PD was fabricated and characterized in this work. The dark *I–V* curve of the junction was found to be exponential, and the junction responded to light, with its photovoltage being a logarithmic function of illuminance. A new phototransistor called the PDOSFET, in which the graphene/n-type poly-Si junction PD was directly integrated with an n-channel MOSFET, was proposed, fabricated, and functionally verified. The channel current of the PDOSFET was efficiently modulated by the photovoltage of the PD and thus responded to light intensity logarithmically. The proposed PDOSFET shows great potential for CMOS image sensor applications, especially for detecting HDR images.

Author Contributions: Conceptualization, K.Y.J.H.; methodology, K.Y.J.H. and P.-W.C.; validation, K.Y.J.H., Y.-Y.T., C.-Y.K., and B.-C.L.; formal analysis, K.Y.J.H., Y.-Y.T., C.-Y.K., and B.-C.L.; investigation, Y.-Y.T., C.-Y.K., and B.-C.L.; resources, K.Y.J.H. and P.-W.C.; data curation, K.Y.J.H.; writing—original draft preparation, K.Y.J.H.; writing—review and editing, K.Y.J.H.; visualization, K.Y.J.H., Y.-Y.T., C.-Y.K., and B.-C.L.; supervision, K.Y.J.H. and P.-W.C.; project administration, K.Y.J.H.; funding acquisition, K.Y.J.H. and P.-W.C. All authors have read and agreed to the published version of the manuscript.

Funding: This research was funded by the Ministry of Science and Technology, Taiwan, R.O.C. under grant numbers MOST 103-2119-M-007-008 -MY3 and MOST 105-2221-E-007-109 -.

Conflicts of Interest: The authors declare no conflict of interest.

References

1. Obradovic, B.; Kotlyar, R.; Heinz, F.; Matagne, P.; Rakshit, T.; Giles, M.D.; Stettler, M.A.; Nikonov, D.E. Analysis of graphene nanoribbons as a channel material for field-effect transistors. *Appl. Phys. Lett.* **2006**, *8*, 142102. [CrossRef]
2. Lemme, M.C.; Echtermeyer, T.J.; Baus, M.; Kurz, H. A graphene field-effect device. *IEEE Electron. Device Lett.* **2007**, *28*, 282–284. [CrossRef]
3. Lin, Y.-M.; Dimitrakopoulos, C.; Jenkins, K.A.; Farmer, D.B.; Chiu, H.-Y.; Grill, A.; Avouris, P. 100-GHz transistors from wafer-scale epitaxial graphene. *Science* **2010**, *327*, 662. [CrossRef] [PubMed]
4. Skulason, H.S.; Nguyen, H.V.; Guermoune, A.; Sridharan, V.; Siaj, M.; Caloz, C.; Szkopek, T. 110 GHz measurement of large-area graphene integrated in low-loss microwave structures. *Appl. Phys. Lett.* **2011**, *99*, 153504. [CrossRef]

5. Xia, F.; Mueller, T.; Lin, Y.-M.; Valdes-Garcia, A.; Avouris, P. Ultrafast graphene photodetector. *Nat. Nanotech.* **2009**, *4*, 839–843. [CrossRef]

6. Mueller, T.; Xia, F.; Avouris, P. Graphene photodetectors for high-speed optical communications. *Nat. Photon.* **2010**, *4*, 297–301. [CrossRef]

7. Xing, F.; Xin, W.; Jiang, W.-S.; Liu, Z.-B.; Tian, J.-G. A general method for large-area and broadband enhancing photoresponsivity in graphene photodetectors. *Appl. Phys. Lett.* **2015**, *107*, 163110. [CrossRef]

8. Bablich, A.; Kataria, S.; Lemme, M.C. Graphene and two-dimensional materials for optoelectronic applications. *Electronics* **2016**, *5*, 13. [CrossRef]

9. Mak, K.F.; Sfeir, M.Y.; Wu, Y.; Lui, C.H.; Misewich, J.A.; Heinz, T.F. Measurement of the optical conductivity of graphene. *Phys. Rev. Lett.* **2008**, *101*, 196405. [CrossRef]

10. Bonaccorso, F.; Sun, Z.; Hasan, T.; Ferrari, A.C. Graphene photonics and optoelectronics. *Nature Photon.* **2010**, *4*, 611–622. [CrossRef]

11. Goykhman, I.; Sassi, U.; Desiatov, B.; Mazurski, N.; Milana, S.; De Fazio, D.; Eiden, A.; Khurgin, J.; Shappir, J.; Levy, U.; et al. On-chip Integrated, silicon-graphene plasmonic Schottky photodetector with high responsivity and avalanche photogain. *Nano Lett.* **2016**, *16*, 3005–3013. [CrossRef] [PubMed]

12. Casalino, M.; Sassi, U.; Goykhman, I.; Eiden, A.; Lidorikis, E.; Milana, S.; De Fazio, D.; Tomarchio, F.; Iodice, M.; Coppola, G.; et al. Vertically illuminated, resonant cavity enhanced, graphene-silicon Schottky photodetectors. *ACS Nano* **2017**, *11*, 10955–10963. [CrossRef] [PubMed]

13. Echtermeyer, T.J.; Britnell, L.; Jasnos, P.K.; Lombardo, A.; Gorbachev, R.V.; Grigorenko, A.N.; Geim, A.K.; Ferrari, A.C.; Novoselov, K.S. Strong plasmonic enhancement of photovoltage in graphene. *Nat. Commun.* **2011**, *2*, 458. [CrossRef] [PubMed]

14. Thongrattanasiri, S.; Koppens, F.H.L.; García de Abajo, F.J. Complete optical absorption in periodically patterned graphene. *Phys. Rev. Lett.* **2012**, *108*, 047401. [CrossRef]

15. Konstantatos, G.; Badioli, M.; Gaudreau, L.; Osmond, J.; Bernechea, M.; de Arquer, P.G.; Gatti, F.; Koppens, F.H. Hybrid graphene-quantum dot phototransistors with ultrahigh gain. *Nat. Nanotechnol.* **2012**, *7*, 363–368. [CrossRef]

16. Pei, Z.; Lai, H.-C.; Wang, J.-Y.; Chiang, W.-H.; Chen, C.-H. High-responsivity and high-sensitivity graphene dots/a-IGZO thin-film phototransistor. *IEEE Elec. Dev. Lett.* **2015**, *36*, 44–46. [CrossRef]

17. De Arco, L.G.; Zhang, Y.; Schlenker, C.W.; Ryu, K.; Thompson, M.E.; Zhou, C. Continuous, highly flexible, and transparent graphene films by chemical vapor deposition for organic photovoltaics. *ACS Nano* **2010**, *4*, 2865–2873. [CrossRef]

18. Wang, X.; Zhi, L.; Müllen, K. Transparent, conductive graphene electrodes for dye-sensitized solar cells. *Nano Lett.* **2007**, *8*, 323–327. [CrossRef]

19. Behura, S.K.; Mahala, P.; Nayak, S.; Jani, O. Graphene as transparent and current spreading electrode in silicon solar cell. *AIP Adv.* **2014**, *4*, 117111. [CrossRef]

20. Yin, Z.; Zhu, J.; He, Q.; Cao, X.; Tan, C.; Chen, H.; Yan, Q.; Zhang, H. Graphene-based materials for solar cell applications. *Adv. Energy Mater.* **2014**, *4*, 1300574. [CrossRef]

21. Fan, G.; Zhu, H.; Wang, K.; Wei, J.; Li, X.; Shu, Q.; Guo, N.; Wu, D. Graphene/Silicon nanowire Schottky junction for enhanced light harvesting. *ACS Appl. Mater. Interfaces* **2011**, *3*, 721–725. [CrossRef] [PubMed]

22. Chen, C.-C.; Aykol, M.; Chang, C.-C.; Levi, A.F.J.; Cronin, S.B. Graphene-silicon Schottky diodes. *Nano Lett.* **2011**, *11*, 1863–1867. [CrossRef] [PubMed]

23. An, Y.; Behnam, A.; Pop, E.; Ural, A. Metal-semiconductor-metal photodetectors based on graphene/p-type silicon Schottky junctions. *Appl. Phys. Lett.* **2013**, *102*, 013110. [CrossRef]

24. Kalita, G.; Hirano, R.; Ayhan, M.E.; Tanemura, M. Fabrication of a Schottky junction diode with direct growth graphene on silicon by a solid phase reaction. *J. Phys. D Appl. Phys.* **2013**, *46*, 455103. [CrossRef]

25. An, X.; Liu, F.; Jung, Y.J.; Kar, S. Tunable graphene–silicon heterojunctions for ultrasensitive photodetection. *Nano Lett.* **2013**, *13*, 909–916. [CrossRef]

26. Liu, X.; Zhang, X.W.; Yin, Z.G.; Meng, J.H.; Gao, H.L.; Zhang, L.Q.; Zhao, Y.J.; Wang, H.L. Enhanced efficiency of graphene-silicon Schottky junction solar cells by doping with Au nanoparticles. *Appl. Phys. Lett.* **2014**, *105*, 183901. [CrossRef]

27. Yim, C.; McEvoy, N.; Duesberg, G.S. Characterization of graphene-silicon Schottky barrier diodes using impedance spectroscopy. *Appl. Phys. Lett.* **2013**, *103*, 193106. [CrossRef]

28. Li, X.; Zhu, M.; Du, M.; Lv, Z.; Zhang, L.; Li, Y.; Yang, Y.; Yang, T.; Li, X.; Wang, K.; et al. High detectivity graphene-silicon heterojunction photodetector. *Small* **2016**, *12*, 595–601. [CrossRef]

29. Riazimehr, S.; Kataria, S.; Bornemann, R.; Bolívar, P.H.; Ruiz, F.J.G.; Engström, O.; Godoy, A.; Lemme, M.C. High photocurrent in gated graphene–silicon hybrid photodiodes. *ACS Photonics* **2017**, *4*, 1506–1514. [CrossRef]

30. Di Bartolomeo, A.; Luongo, G.; Iemmo, L.; Urban, F.; Giubileo, F. Graphene-silicon Schottky diodes for photodetection. *IEEE Trans. Nanotechnol.* **2018**, *17*, 1133–1137. [CrossRef]

31. Wang, Y.; Zhang, Z.; Zhang, Y.; Feng, M.; Tsang, H.K. Low power saturation of waveguide-integrated graphene-silicon Schottky diode on micro-ring resonator. In Proceedings of the 2019 Asia Communications and Photonics Conference (ACP), Chengdu, China, 2–5 November 2019.

32. Zeng, L.-H.; Wang, M.-Z.; Hu, H.; Nie, B.; Yu, Y.-Q.; Wu, C.-Y.; Wang, L.; Hu, J.-G.; Xie, C.; Liang, F.-X.; et al. Monolayer graphene/germanium Schottky junction as high-performance self-driven infrared light photodetector. *ACS Appl. Mater. Interfaces* **2013**, *5*, 9362–9366. [CrossRef] [PubMed]

33. Lin, J.-H.; Lin, Y.-J.; Chang, H.-C. Tuning electrical parameters of graphene/p-type polycrystalline silicon Schottky diodes by ultraviolet irradiation. *Appl. Phys. A* **2015**, *118*, 361–366. [CrossRef]

34. Laghla, Y.; Scheid, E. Optical study of undoped, B or P-doped polysilicon. *Thin Solid Films* **1997**, *306*, 67–73. [CrossRef]

35. Li, X.; Cai, W.; An, J.; Kim, S.; Nah, J.; Yang, D.; Piner, R.; Velamakanni, A.; Jung, I.; Tutuc, E.; et al. Large-area synthesis of high-quality and uniform graphene films on copper foils. *Science* **2009**, *324*, 1312–1314. [CrossRef] [PubMed]

36. Lin, Y.-C.; Jin, C.; Lee, J.-C.; Jen, S.-F.; Suenaga, K.; Chiu, P.-W. Clean transfer of graphene for isolation and suspension. *ACS Nano* **2011**, *5*, 2362–2368. [CrossRef] [PubMed]

37. Kwon, Y.-J.; Lee, C.-J.; Kim, D.-K.; Hahm, S.-H. Dual-band responsivity of AlGaN/GaN MSM UV photodiode. In Proceedings of the 2012 IEEE International Conference of Electron Devices and Solid-State Circuits (EDSSC), Bangkok, Thailand, 3–5 December 2012; p. 2.

38. Sze, S.M.; Coleman, D.J.; Loya, A. Current transport in metal-semiconductor-metal (MSM) structures. *Solid-State Electron.* **1971**, *14*, 1209–1218. [CrossRef]

39. Wuu, S.-G.; Chien, H.-C.; Yaung, D.-N.; Tseng, C.-H.; Wang, C.S.; Chang, C.-K.; Hsiao, Y.-K. A high performance active pixel sensor with 0.18 um CMOS color imager technology. In Proceedings of the International Electron Devices Meeting, Washington, DC, USA, 2–5 December 2001; Technical Digest. IEEE: Piscataway, NJ, USA, 2001; pp. 555–558.

40. Chien, H.-C.; Wuu, S.-G.; Yaung, D.-N.; Tseng, C.-H.; Lin, J.-S.; Wang, C.S.; Chang, C.-K.; Hsiao, Y.-K. Active pixel image sensor scale down in 0.18 um CMOS technology. In Proceedings of the International Electron Devices Meeting, San Francisco, CA, USA, 8–11 December 2002; Technical Digest. IEEE: Piscataway, NJ, USA, 2002; pp. 813–816.

41. Yoon, K.; Kim, C.; Lee, B.; Lee, D. Single-chip CMOS image sensor for mobile applications. *IEEE J. Solid-State Circuit* **2002**, *37*, 1839–1845. [CrossRef]

Review

Integrated Photodetectors Based on Group IV and Colloidal Semiconductors: Current State of Affairs

Principia Dardano * and Maria Antonietta Ferrara *

National Research Council (CNR), Institute of Applied Sciences and Intelligent Systems,
Via Pietro Castellino 111, 80131 Naples, Italy
* Correspondence: principia.dardano@na.isasi.cnr.it (P.D.); antonella.ferrara@na.isasi.cnr.it (M.A.F.)

Received: 10 August 2020; Accepted: 5 September 2020; Published: 8 September 2020

Abstract: With the aim to take advantage from the existing technologies in microelectronics, photodetectors should be realized with materials compatible with them ensuring, at the same time, good performance. Although great efforts are made to search for new materials that can enhance performance, photodetector (PD) based on them results often expensive and difficult to integrate with standard technologies for microelectronics. For this reason, the group IV semiconductors, which are currently the main materials for electronic and optoelectronic devices fabrication, are here reviewed for their applications in light sensing. Moreover, as new materials compatible with existing manufacturing technologies, PD based on colloidal semiconductor are revised. This work is particularly focused on developments in this area over the past 5–10 years, thus drawing a line for future research.

Keywords: photodetector; semiconductor; microphotonics; group IV; colloidal systems

1. Introduction

Microphotonics is a branch of microelectronics where devices manipulates light on a microscopic scale and it is used in optical networking. Essentially, it consists in wafer-level integrated devices and systems that allow to emit, transmit, detect, and process light along with other forms of radiant energy with photon as the quantum unit [1]. The immunity of optical signals to external influences, such as the intensification of light beams without interaction and the possibility to realize cheap densely packed components capable of large-scale integration, make microphotonics an area of great interests in research and development [2].

In order to exploit the existing technology in microelectronics, Silicon (Si) is considered the material to be investigated to realize integrated photonic circuits. For this reason, silicon-based microphotonics has gained growing interest over the past decades. Indeed, fabrication technologies employed for micro- and nano-silicon photonics allow the integration of electronic, photonic, and sensing devices on the same chip with a very low cost. In recent decades, considerable efforts have been devoted to the development of new silicon photonic components that have led to innovative solutions with applications in the field of telecommunications and multichip optical interconnections and which promise to improve the performance of the next generation of commercial processors.

Such efforts have given rise, for example, to new complementary metal-oxide-semiconductor (CMOS) configurations of optical modulators capable of operating in the gigahertz regime, waveguides and photonic crystal switches (PhC) to direct the light on chips on a sub-micrometric scale [3]. These include the innovative use of nanostructures with a negative effective refractive index capable of guiding light with very low losses [4,5] and switching its optical path through unprecedented effects [6]. Furthermore, the matching of processing technologies have paved the way for new devices integrated with fluidic, electronic, and photonic circuits for sensing detection and healthcare, which have led to the development of a new generation of biochips to be applied in fields ranging from biomedical to environmental [7–9].

On the other hand, for a complete manipulation of the light on a single intelligent microsystem, close integration with an emissive source and a detector capable of receiving the modulated light and switching it into an electrical signal is crucial. In this direction, new experiments on Raman emission in silicon nanocrystals have shown an extremely high increase of the optical gain [10–12], encouraging the research on this way.

Finally, to complete a perfectly integrated optoelectronic multisystem, it is necessary to comply with the need to detect light and convert it into an electrical signal, all using semiconductor basic materials, such as Silicon, and metals compatible with current fabrication technologies. In this sense, considerable efforts have been made to obtain an efficient photodetector that can be easily integrated and processed through standard electronic manufacturing steps.

Basically, photodetector (PD) is a device that converts optical signals into electrical signals, thus it is also known as O/E convertor. The most used types of photodetector in optical communication systems are semiconductor-based photodetectors, usually called as photodiodes, because of their high detection efficiency, fast detection speed, and small size [13]. Photodiodes, like the structures of laser diodes, are based on the p-n junctions or p-i-n junction, where the diode is integrated with a wide, undoped intrinsic semiconductor region between a p-type semiconductor and an n-type semiconductor region. This intrinsic region enhances detection speed.

PD can be used in different configurations for different purposes:

- Single light sensors: A single sensor is used to detect overall light levels, it is useful to quantify the total optical power or light intensity;
- 1-D array light sensor: A line of PD is used to quantify the distribution of optical power or light intensity along a line. Combined with a wavelength splitter it can be used in a spectrophotometer; and
- 2-D array light sensor: A NxM matrix of photodetectors can be used to form images with NxM resolution.

Photodetectors are built by semiconductor materials which are able to absorb emitted photons and generate electron–hole pairs under the light excitation. A typical configuration is made of inorganic semiconductors [14], for example, GaN-based photodetectors are used for the ultraviolet sub-band (0.25–0.4 µm), Si-based photodetectors are implemented for the visible sub-band (0.45–0.8 µm), and InGaAs-based photodetectors are developed for near-infrared (NIR) sub-band (0.9–1.7 µm). An interesting and explicative graph that summarizes composition, energy gap and wavelength for most used material systems for infrared (IR) photodetectors is reported in Figure 1 [15], while some physical and optical properties at room temperature of narrow gap semiconductors are reported in Table 1 (Energy gap, E_g, quantum efficiency, n_i, dielectric permittivity, ε, and magnetic permeability, μ_e and μ_h).

Table 1. Some physical and optical properties at room temperature of narrow gap semiconductors [15].

Material	E_g (eV)	n_i (cm^{-3})	ε	μ_e (10^4 cm^2/Vs)	μ_h (10^4 cm^2/Vs)
InAs	0.359	9.3×10^{14}	14.5	3	0.02
InSb	0.18	1.9×10^{16}	17.9	8	0.08
PbS	0.42	1.0×10^{15}	172	0.05	0.06
PbSe	0.28	2.0×10^{16}	227	0.10	0.10
PbTe	0.31	1.5×10^{16}	428	0.17	0.08
In$_{0.53}$Ga$_{0.47}$As	0.75	5.4×10^{11}	14.6	1.38	0.05
Pb$_{0.44}$Sn$_{0.56}$Te	0.1	2.0×10^{16}	400	0.12	0.08
Hg$_{1-x}$Cd$_x$Te	0.07–0.25	$(0.23–2.3) \times 10^{16}$	16.7–18.0	0.6–1.0	0.01

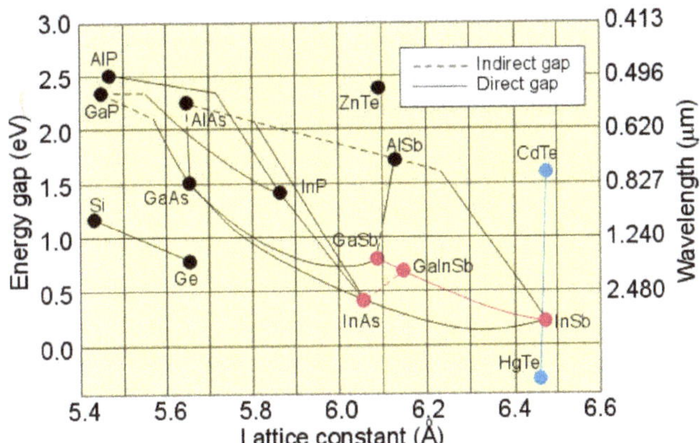

Figure 1. Composition and wavelength diagram of semiconductor material systems (Reprinted from [Rogalski, A.; Antoszewski, J.; Faraone, L. Third-generation infrared photodetector arrays. *J. Appl. Phys.* **2009**, *105*, 091101, doi:10.1063/1.3099572], with the permission of AIP Publishing).

While the InSb has narrow band gaps in the mid-wave IR (MWIR), where it dominates as a semiconductor for PDs, PDs based on $Hg_{1-x}Cd_xTe$ (MCT) have the great advantage of being able to tune the wavelength of operation from MWIR to long-wave IR (LWIR), adjusting the composition.

As it is known, semiconductor materials have a crystalline solid structure, but to have a device with optimal performance, this structure must not have defects and it should be single crystal. To achieve this, crystals must be grown by coupling the crystal lattice to that of the substrate. This is achieved either with molecular beam epitaxy (MBE) or chemical vapor deposition (CVD) and involves high costs for industrial production. Using heterostructured semiconductors, whose molecular structure has a broadband super-lattice, it was possible to obtain infrared absorption with intersubband transitions within one of the wells of the superlattice (GaAs/AlGaAs). PD based on this effect are called Quantum Well Photodetectors (QWP) [16] or Strained Layer Super-lattice Photodetectors (SLSP). Another approach has been the development of infrared quantum dot detectors (QDIP) based on self-assembled epitaxial quantum dots [17–21]. However, initially the use of QD in photodetectors presented two main difficulties: Limited control of QD size and low density of QD grown epitaxially. To date, InSb and MCT are the most used materials for the production of commercial quantum detectors [22]. Such quantum detectors exhibit high performance with high efficiency and high speed. However, the production of these detectors continues to be very expensive. In addition, considering their low operating temperature, they require a cooling system, so their use is limited to defense and astronomical research.

There is a plenty of materials studied in different configurations to realize an efficient photodetector, however most of them show complexity, high cost and difficulty to integrate with the standard CMOS process technologies. The research of alternative solutions that will address these problems requires a new active materials development or, in alternative, exploits the already known materials generally used for opto-electronic integration. Significant progress has been made in the past decade, overall using nanocrystal formations and colloidal quantum dot for IR sensing developments [23,24].

For these reasons, we focus our review paper on the current state of photodetectors based on group IV and colloidal semiconductors, which allow high compatibility with CMOS process and good performance.

2. Figure of Merits for Characterizing Photodetectors

In order to characterize the performance of photodetectors and compare different kinds of PD, some key figure-of-merit parameters are typically used. The main ones are summarized below [25,26]:

- *Quantum efficiency (QE)*: is the number of carriers (electrons or holes) generated per photon of a given energy. There are two types of QE: Internal quantum efficiency (IQE) that represents the number of charge carriers collected by the PD to the number of absorbed photons of a given energy, and external quantum efficiency (EQE) that is the number of charge carriers collected by the PD to the number of incident photons of a given energy.
- *Responsivity (R)*: The ratio of photogenerated current I_{ph} to input light power P_{in}, indicating the electrical response of an optical signal in units of A W^{-1},

$$R = \frac{I_{ph}}{P_{in}} \tag{1}$$

- *Spectral response*: Describes the responsivity of a PD as a function of photon frequency.
- *Specific detectivity (D*)*: Represents the ability to detect weak optical signals. D^* depends on the specific measurement conditions comprising the bias voltage, operating temperature, wavelength and modulation frequency. It is expressed in units of cm Hz$^{1/2}$ W^{-1} (Jones),

$$D^* = \frac{R(A\Delta f)^{1/2}}{i_n} \tag{2}$$

where A is the effective area of the device, Δf is the electrical bandwidth and i_n is the noise current that includes the shot noise from the dark current, Johnson noise and flicker noise. However, D^* is generally considered in the shot noise limit, thus the Johnson noise and flicker noise can be neglected. Thus, Equation (2) can be simplified as $D^* = RA^{1/2}/(2eI_{dark})^{1/2}$, where e is the electron charge, and I_{dark} is the dark current.

- *Response time*: The time needed for a PD to rise or fall (τ_r/τ_f) from 10% to 90% or 90% to 10% of the final output.
- *Noise-equivalent power (NEP)*: Indicates the lowest amount of light power needed to generate a signal comparable to the noise of the device (i.e., a signal-to-noise ratio of $\approx$1) when the electrical bandwidth of the noise measurement is equal to 1 Hz. The NEP (W Hz$^{-1/2}$) is proportional to the reciprocal of D^*,

$$NEP = \frac{(A\Delta f)^{1/2}}{D^*} = \frac{i_n}{R} \tag{3}$$

- *−3 dB bandwidth (f_{-3dB})*: Defined by the incident light modulation frequency at which the output signal is half-attenuated respect to its value under continuous wave illumination.
- *Linear dynamic range (LDR)*: The range of incident light for which the detector responds linearly.
- *Fill factor*: The ratio of a light sensitive area of a pixel to its total area, typically it characterizes an image sensor array. The effective fill factor can be increased, often to nearly 100%, by using microlenses.

In an ideal photodiode, each photon of an input optical signal is immediately converted into a free electron and the generated current is linearly proportional to the input optical power. However, in real photodiodes, a lower absorption of photons and a leakage in carrier collection into the semiconductor material cause a non-perfect conversion in electrons of incoming photons. Moreover, the electric structures show an equivalent parasitic RC and the carrier transient effect limits the photodetection speed. Thus, the detection speed and the responsivity of a photodiode depend on the material quality, the bandgap structure of the semiconductor, the design of the device photonic and the electrodes [13].

Also, light reflection effect on the photodetector surface produces a leakage into the transduction, thus the PD, or the array of PD, is typically covered by an illumination window with an anti-reflection coating, reducing this effect. Furthermore, in an actual photodiode shot noise, thermal noise, and dark-current noise generates a signal-to-noise ratio (SNR) degradation in the photodetection process, which incisively affects the performance of an optical communication system. By increasing the load resistance, thermal noise can be decreased. Also, improvement of material quality and junction structure optimization can reduce the reverse saturation current and, then, dark-current noise. Nevertheless, shot noise is a white noise generated by the photo detection process itself and cannot be reduced. It is called the quantum limit of the optical system and it represents the fundamental limit of the photodiode performance.

Often, photodetectors are classified on the basis of the mechanism involved in the detection. As well know, the interaction between light and matter strongly depends by the photon energy and the band structure of the material constituent the photodetector. Then, effects of the light-matter interaction exploited for the photon-carrier transduction in PDs are summarized as following [25–27]:

- *Photoemission or photoelectric effect*: Energy of photons supplies exactly the energy gap from the conduction band to free electrons, increasing the mobility of electrons.
- *Thermal effect*: Energy of photons supplies to mid-gap transition states then an electron decay back to lower bands, generating phonon and thus heat.
- *Photochemical effect*: In some materials, photons can induce a chemical change as crosslinking or the destruction of a chemical bond.
- *Polarization effect*: In some materials, photons can cause changes in polarization states, which can change the refractive index or induces birefringence effects.
- *Weak interaction effects*: Photons can lead to secondary effects such as gas pressure changes or photon drag [28,29].

3. Group IV Semiconductors

The main semiconductors used to realize a photodetector, are summarized in the table of elements reported in Figure 2; in particular, the group IV semiconductors are currently the main materials for electronic and optoelectronic devices fabrication. Historically, efficient detection is provided by exploiting the direct bandgap of all III–V semiconductors; however, the integration of materials outer to group IV semiconductor with Si integrated circuit (IC) is complicated [30]. As example, in an integrated system the PD has to be closely connected to the biasing system and the electronic amplifying circuits [31], in order to exhibit the best possible performance.

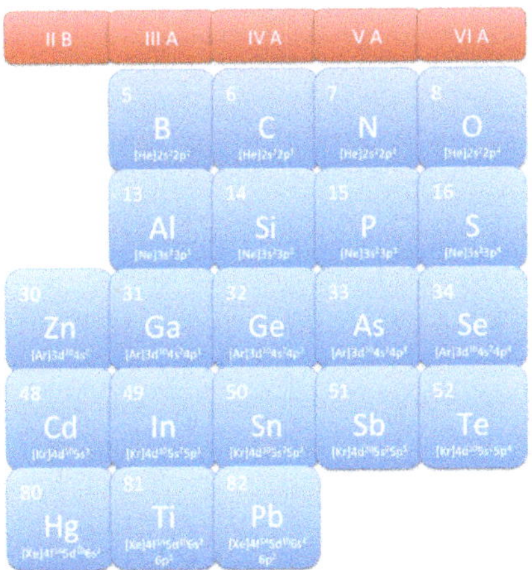

Figure 2. Table of elements involved in semiconductor based photodetectors (PDs) production.

The compatibility with Si technology of the IV Group semiconductors together with the efficient near-infrared light detection, enable fabrication of PDs and Si CMOS circuits in a simultaneously and monolithically integrated way [32]. Among PD systems, metal-semiconductor-metal (MSM) structures, with group IV semiconductor, are promising candidates for sensor optoelectronic integrated circuits (OEICs) due to the low detector capacitance, large device bandwidth, internal gain and, obviously, easiness of integration with electronic circuits. In spite of that, a lower bandgap increases the dark current associated with a low Schottky barrier in MSMs where the semiconductor is Ge or Si. As reported before, in order to limits the dark current effect, a reverse bias has to be applied leading to an extra power consumption. The connected heat generation increases the already high temperature of the Si IC substrate that absorbs the heat of the whole device. In particular, in Ref. [33] authors show both theoretically and experimentally a significant suppression of MSM-PD dark current by applying asymmetric metal electrodes and by modifying the Schottky barrier heights.

A similar technique has also been applied for MSM III-V PDs [34]. On the other hand, the only use of a wider bandgap semiconductor layer within the metal-semiconductor contacts can increase the Schottky barrier in MSMs and, thus, enhance the dark current effect [35].

Bulk silicon is naturally important for its main role in integrated circuits, but substantial challenges also derive from other elements by the group IV bulk materials and their alloys, nanocomposites, nanostructures, films of different thickness, and heterostructures. Advances in device performance are obtained by means of new defect engineering production, novel growth techniques, and improvements in diagnostic tools.

The advantage to consider group IV semiconductors is that they are widely studied thus modeling of defect generation, modeling of crystal growth, growth of group IV alloy crystals, low quality polycrystalline silicon refinement, including control of dopants and wafering technologies, and defect evolution in wafering processes are all well-known issues. Moreover, nanostructures of/on group IV semiconductors are largely developed, as well as modeling and simulation of epitaxial structures, heterogeneous integration of Si or Ge with III-V epitaxial device quality layers, growth of 2D materials (e.g., graphene, silicene, and germanene) on silicon, deposition of amorphous and crystalline thin layers and silicon membranes. Fundamental research on point defects and extended defects in group IV semiconductors is continuously in progress, as well as the study of dislocation engineering by substrate and process optimization. Additionally, considering the several technological applications for group IV semiconductors, such as silicon on insulator (SOI) devices, high speed and high frequency electronic devices, photonics and light emitting devices, and power devices, it is natural to invest resources to enhance the performance of devices based on these semiconductors.

3.1. Si

Although in the visible spectrum the PDs in Si have reached the phase of commercial development, in NIR band the use of Si for the realization of PD is difficult due to its transparency at wavelengths greater than 1.1 µm. Generally, the integration of germanium (Ge) in a SiGe heterostructure is used to overcome this drawback [36,37], however a reticular superlattice mismatch of 4.3% leads to a very high leakage current, affecting the performance of the device. A two-step epitaxial growth technique can mitigate these effects [36,37] but, unfortunately, does not completely remove the defect center responsible for the high leakage current. Furthermore, the grown buffer layer gives rise to thermal and flatness problems [38], which prevent the possibility of a monolithic integration of Ge on Si. Additionally, in p-i-n structures the Ge needs a very thick intrinsic layer (especially with respect to the gallium arsenide (InGaAs)) because at 1550 nm the Ge shows a lower absorption.

In this paragraph, we present a brief overview on all-Si photodetectors at NIR wavelengths. In these PDs, the most used physical effects, that allow the absorption at sub-band wavelength, are the internal photoemission absorption (IPA) in presence of positive charged particles or molecules, two photon absorption (TPA), mid-bandgap absorption (MBA), and surface state absorption (SSA). A quantitative comparison of PDs in the literature, divided by absorption effects are shown in Table 2.

Table 2. Some physical and optical properties at room temperature of narrow gap semiconductors.

Responsivity (A/W)	V_{bias} (V)	Operating Wavelength (nm)	Dark Current/Leakage	Effect	Ref.
4.6×10^{-3}	-1	1550	3 nA	IPA NiSi$_2$/p-Si Schottky barrier	[39]
8×10^{-3}	-1	1550	~3 nA	MBA Proton implantation	[40]
0.8×10^{-3}	-0.1	1550	6 µA	IPA Surface plasmon polariton	[41]
64×10^{-3}	-20	1440	0.1 µA	MBA He^{2+} implantation	[42]
0.1	-2	1549	0.1 nA	MBA Si$^+$ implantation	[43]
8×10^{-6}	$-0,1$	1550	-	IPA Cu/p-Si Schottky barrier	[44]
0.08×10^{-3}	-1	1550	10 nA	IPA Cu/p-Si Schottky barrier	[45]
0.5–0.8	-5	1550	2.5 nA/mm	MBA Si$^+$ implantation	[46,47]
50×10^{-3}	-0.5	1330	120 µA/cm^2	MBA Laser irradiation in presence of SF$_6$	[48]
36×10^{-3}	-11	1575	0.12 µA	SSA	[49]
0.25×10^{-3}	-15	1541.5	2.5 nA	SSA-ring resonator	[50]
6×10^{-3}	-3	1550	15 pA	TPA Photonic crystal resonators	[51]
2×10^{-6}	1	1300	-	TPA Hemispherical structure	[52]
0.18	-6	1550	5 µA	IPA Al-porous Si Schottky barrier	[53]
35×10^{-3}	-0.5	1550	120 µA/cm^2	MBA in presence of SF$_6$	[54]

To improve the Si based photodetectors performance, many approaches have been considered and proposed. Among them, encouraging results have been obtained by exploiting the internal photo absorption effect (IPA), that is, the use of photon-assisted transmission of hot carriers across a potential barrier at metal-semiconductor interfaces. Several new structures have been reported in literature combining, for example, IPA with nanoscale metallic structures, comprising Si nanoparticles (NPs) [39], metal stripes allowing surface plasmon polaritons (SPPs) [40,41], metallic gratings [42] and new structures based on two-dimensional materials (like graphene) capable to substitute metal in the Schottky junction [43]. Moreover, as reported in some complete review on this topic [44,45], due to the unipolar nature of the Schottky junction, IPE based PDs are very fast and can be monolithically integrated with Si-based CCD for IR applications [46].

A 50 Gb/s pure silicon waveguide photodetector at 1310 nm datacom wavelength has been also developed [47]. Its working principle is based on a combination of TPA, SSA, and photon assisted tunneling (PAT) effects in a Si PN junction; responsivity was found to be 0.6 A/W, dark current was of 850 nA at 5.84 V reverse bias and eye SNR was of 7.3.

Nanostructured photodetectors have attracted a lot of interests in the recent past, due to their large surface-to-volume ratios and the reduced dimension, that lead to superior performances in terms of responsivity and photoresponse gain with respect to bulk crystals. Indeed, the large surface areas of nanostructures gives rise to a prolonged photocarrier lifetime leading to higher sensitivity and responsivity. Moreover, nanostructured photodetectors show a fast response speed owing the low dimensionality which shortens the transit time during photoelectric processes [48–50], and good compatibility with semiconductor technology. On this line of research, 1D inorganic nano-wires (NWs) PD have demonstrated high responsivity, fast response, specific detectivity, high spectral selection, good flexibility, and low energy consumption [51]. The first avalanche photodetectors (APD)-based nanoscale p–i–n junction Si nanowire (SiNW) was demonstrated in 2006 and this device showed an avalanche breakdown mechanism with large reverse bias [52]. After that, several studies were performed on Schottky junction NIR photodiodes and nano-heterojunction NIR-PDs based on SiNW [53] combined with metal films (e.g., Cu [54], Ag [55] and Au/Cr [56]), graphene oxide [57], and other nanostructured semiconductor comprising carbon quantum dots [58].

With the aim to improve the excitation efficiency, Yatsui et al. [59] investigated optical near-field excitation that is confined in a nano-scale. Here, due to the uncertainty principle, the interband transitions between different wave numbers are excited; therefore, optical near-field can directly excite the carrier in the Si indirect bandgap. The authors developed a lateral Si p–n junction with Au nanoparticles as sources to generate the field confinement and they demonstrated that the photo-sensitivity rate increases of 47.0%. Moreover, the far-field excitation is eliminated when the thin lateral p–n junction is used, confirming a 42.3% increase in the photosensitivity rate [59].

Even though the Si-based PD limitation can be partially solved by femtosecond laser processing, the surface defects and carrier activation rates produces a large dark current and narrow spectral response, which are drawbacks. To overcome this, Huang et al. [60] have recently proposed a rapid thermal annealing and hydrogenated surface passivation demonstrating, at optimal conditions, a sub-bandgap responsivity of 0.80 A W^{-1} for 1550 nm at 20 V at room temperature; these results are comparable with commercial Ge based PDs and higher than previously reported Si photodiodes. This Si-based photodetector shows a spectral range of 400 ÷ 1600 nm and a competitive detectivity (1.22 × 10^{14} Jones at −5 V). Finally, its responsivity was 1097.60 A W^{-1} for 1080 nm at 20 V, which is the highest reported to date in Si photodetectors, thus opening good prospects for black silicon in IR light detection, night vision imaging, and fiber-optic communication [60].

3.2. Ge

Germanium has a direct bandgap energy of 0.8 eV, high optical absorption over the 1.3–1.55 μm wavelength range used for fiber-optic communications [61] and a perfect compatibility with the conventional silicon processing technology. For these characteristics, Ge-based photodetectors have been widely studied since the end of 1990s [62–65] and, currently, Ge PDs performance are comparable to that of III-V materials.

For nanophotonics applications, Ge photodetectors are often integrated at the end of optical waveguides, indeed in this configurations light absorption takes place along the optical mode propagation direction and perpendicularly to the carrier collection path. Moreover, both Ge homo-junction [62–64] and Si-Ge-Si hetero-junction [65–67] photodetector architectures, which are made with a p-i-n junction where light absorption occurs in the intrinsic regions, have been experimentally demonstrated. A lateral hetero-structured silicon-Ge-silicon (Si-Ge-Si) junction was realized to obtain a p-i-n waveguide photodetectors working under low reverse bias at 1.55 μm [67]. Such hetero-structured Si-Ge-Si photodetector allows a superior signal detection of high-speed data traffic, having efficiency-bandwidth products of ~9 GHz at −1 V and ~30 GHz at −3 V, with a leakage dark current as low as ~150 nA. For conventional 10 Gbps, 20 Gbps, and 25 Gbps data rates, a bit-error rate of 10^{-9} has been obtained, leading to optical power sensitivities of −13.85 dBm, −12.70 dBm, and −11.25 dBm, respectively. An interesting solution for the Ge-on-Si PD was proposed by Tzu et al. [68]; the authors developed arrayed germanium-on-silicon waveguide photodetectors for high-power analog applications. Arrays of photodetector with 2, 4, and 8 elements allow to have output powers of −0.4 dBm, 10 dBm, and 14.3 dBm at 18 GHz, 12 GHz, and 5 GHz, respectively. The 4-photodiode array shows a 5-dB improvement over a single PD. An integrated 20 GHz receiver based on a couple of balanced PDs and a Mach-Zehnder delay line interferometer has been also demonstrated in an optical phase-modulated link.

Recently, several solution-processed has been proposed in order to obtain good performance of PDs based on Ge. For example, Hu et al. successfully fabricated a perovskite/germanium heterojunction photodetector with excellent photo-response properties [69]. The heterostructure device was realized by a uniform and pinhole-free perovskite film deposited on top of a single-crystal germanium layer, as illustrated in Figure 3a. Results demonstrated a broad spectrum compared with the single-material-based device, as showed in Figure 3b where the photon response properties are characterized from the visible to near-infrared spectrum. In particular, at optical fiber communication wavelength of 1550 nm this device presents the highest responsivity of 1.4 A/W, while at a visible light wavelength of 680 nm, it exhibits considerable detectivity and responsivity of 1.6×10^{10} Jones and 228 A/W, respectively. The authors have been also estimated the photoconductive gain of the realized device. The gain is defined as [69]

$$G = \frac{I_{ph}}{I_{PI}} = \frac{\tau}{t_p} = \frac{\tau \xi (\mu_e + \mu_h)}{L} \tag{4}$$

where I_{PI} is the primary photocurrent, τ is the carrier lifetime, t_p is the carrier transit time through the electrodes, ξ is the applied electric field, L is the width of the channel, μ_h and μ_e are the hole and electron mobility, respectively. Some of the photogenerated electrons are moved from the perovskite layer to the germanium layer when visible light excitation are considered. Respect to the perovskite, the long carrier lifetime (≈ 200 μs) and ultra-high electron mobility (≈ 3800 cm^2 V^{-1} s^{-1}) in the germanium layer lead to an enhanced gain of about 10^4 in this heterojunction photodetector.

In 2018, a PDs based on Ge QDs fabricated on Ge substrates has been demonstrated with a broadband photoresponse spectral range ($\lambda = 400$–1550 nm) [70]. Its responsivity at room temperature was up to 1.12 A/W, the obtained internal quantum efficiency was IQE = 313%, higher than conventional Si and Ge photodiodes, specific detectivity at room-temperature was D* $\approx 210^{10}$ cm Hz$^{1/2}$ W^{-1} both at visible $\lambda = 640$ nm and telecom $\lambda = 1550$ nm wavelengths. When the operating temperature and incident power were decreased, sharply improved performance were obtained, indeed at T = 100K for an incident power of 10 nW at $\lambda = 1550$ nm, the device showed D* = 1.110^{12} cm Hz$^{1/2}$ W^{-1} and IQE = 1000% [70].

In a very recent work, MSM NIR photodiodes are fabricated on the epitaxial uniform Ge film with a grain size of up to 12 μm, with two diverse electrodes and two different surface passivation interlayers [71]. When electrodes were made of amorphous-Ge (30 nm)/Al (100 nm) and TiO$_2$ (5 nm)/

Au (80 nm), MSM devices show an average spectral responsivity of 0.50 ± 0.16 A/Wand 0.35 ± 0.09 A/W at 1550 nm and voltage bias of -3V, respectively. The maximum spectral responsivity reported was of 0.78 A/W and 0.48 A/W, respectively.

Considering the recent encouraging reported studies and results, together with a well-tested compatibility with the conventional CMOS technology, Ge can be considered a material on which to continue investigating.

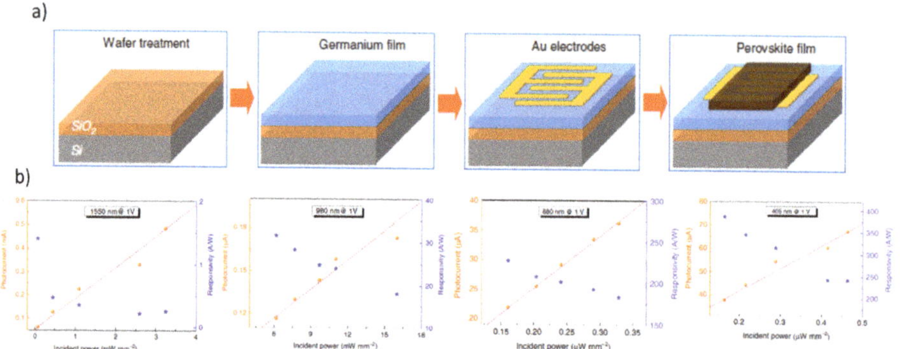

Figure 3. (**a**) Schematic illustration of the germanium/perovskite heterojunction device fabrication process. From left to right: a cleaned SiO_2/Si substrate, growth of a germanium layer by molecular beam epitaxy (MBE), Au electrode deposition on the substrate, perovskite layer construction by the vapor-solution method. (**b**) I_{ph} and R values under an illumination wavelength of 1550 nm, 980 nm, 680 nm and 405 nm, from left to right, respectively (adapted from [69]).

3.3. Carbon

The carbon family comprises fullerene [72], carbon nanodots (CQDs) [73], carbon nanotubes (CNTs) [74,75], graphene [76], and graphene quantum dots (GQDs) [77]. Considering the superior and uniquely chemical, optical, physical, mechanical, and electronic properties of these innovative materials [78], they have attracted a growing research interest in the last couple of decades.

In particular, CNTs and graphene have shown unique potential for a broad range of photodetectors from the ultraviolet to the THz [79]. In this section we introduce the basic properties of both CNTs and graphene based photodetectors that have been realized in the last few years. The performance status of these detectors is reported, and their potential for further improvement is discussed.

3.3.1. Carbon Nanotube

CNTs are tubes made of carbon with diameters in the order of nanometers. When diameters are in the range of a nanometer, carbon nanotubes are called single-wall carbon nanotubes (SWCNTs). Respect to devices made with other semiconductors materials, CNT based photodetectors show a fine IR detection together with low fabrication cost, ease of production and scalability, actually making CNTs highly promising nanomaterials for multiwavelength, room-temperature IR detection applications [80]. Up to now, several IR photodetectors based on both individual CNT and CNTs films [81,82] have been successfully proven. Moreover, few years ago, a prototype of infrared camera was realized based on single CNT photodetectors [83].

In order to develop an efficient CNT photodiodes, based on photovoltaic mechanism, an efficient separation electron–hole pairs into free carriers should be obtained as well as an increase of the absorption cross-section of the device. In particular, the first studies on the realization of CNT photodiodes demonstrated that electron–hole pairs separation and photocurrent generation can be obtained by the built-in field in a CNT Schottky and p-n diode [84–86]. Moreover, even if CNT photodiode attracts a lot of research interest to study the mechanisms of photocurrent generation along with many other optoelectronic properties [86], the extremely small absorption cross-section of individual CNT devices leads to a low sensitivity [87]. Nevertheless, significant improvements have been reached in the recent past, allowing to increase the responsivity from ~pA/W to ~A/W [85,88].

Since the pioneering work proposed by Lee et al. in 2005 [85], where the first realization of individual semiconducting SWCNT p-i-n diode was demonstrated, several other studies were performed on these kind of devices. In detail, the device realized by Lee et al. consisted in a SWCNT where two split gates where formed in adjacent parts by using electrostatic doping to realize n-region and p region, respectively. The photocurrent generated under IR illumination with a power density ~20 W/cm^2 was of ~5 pA, whereas the fill factor (FF) and the efficiency η were of 0.33–0.52 and ~0.2%, respectively.

In 2011 a photodetector based on SWCNT/C(60) in heterojunction photovoltaic was demonstrated with an EQE of 12% and a detectivity approaching ~10^{12} cm Hz$^{1/2}$ giving a proposal for the development of the next-generation high-performance solar cells based on CNTs [89]. Thin films of highly purified semiconducting CNTs were used as NIR optical absorbers in heterojunction photodetector and photovoltaic devices with the electron acceptor C(60), leading to a 10-fold enhanced gain in zero-bias quantum efficiency and significant gains in power conversion efficiency respect to the implementations of more electrically heterogeneous CNT/C(60) devices. In particular, the exciton migration along an effective length scale in the nanotube films has been related to the device efficiency, validating the high IQE through photoluminescence quenching, and demonstrating that for diameters <1.0 nm the driving force for exciton dissociation at the fullerene-fullerene heterointerface is optimized [89].

Liang et al. demonstrated a ~6-fold higher optical absorption in a single-tube diode photodetector monolithically integrated with a Fabry–Pérot microcavity, due to the confined effect of the designed optical mode [90]. The obtained photodetector shows a spectral FWHM of about 33 nm at a signal wavelength of 1200 nm and a higher suppression ratio until a power density of 0.07 W cm^{-2} and photocurrent of 5 pA. Moreover, taking advantage by the theory of the "resonance and off-resonance" cavity, the authors realized cavity-integrated chirality-sorted CNT-film detectors for specific target signal detection, operating at zero bias and resonance-allowed mode. By integrating multiple array detectors on a chip working at different target signals, a multiwavelength signal detector system can be obtained; this device can be used in the fields of colour imaging, monitoring, signal capture, biosensing, and on-chip or space information transfers [90].

A broadband nano-photodetector based on single-layer graphene-CNT thin film (SLG-CNTF) Schottky junction was studied and developed by Zhang et al. [80]. SLG-CNTF device shows peak sensitivity at 600 and 920 nm, a fast response speed (see Figure 4a; τ_r = 68 μs, τ_f = 78 μs, much faster than several seconds response time of the graphite quantum dots/graphene heterojunction [91]) and good reproducibility for switching frequencies in the range 50–5400 Hz (the normalized photocurrent at different frequencies shown in Figure 4b). The responsivity and detectivity at different bias voltages were investigated, as shown in Figure 4c, where is evident that both parameters increase with decreasing bias voltage. Moreover, the authors studied the photocurrent at different light intensities (from 0.25 to 12 mW cm^{-2}) finding a high dependence of the photocurrent of the SLG-CNTF Schottky junction photodetector on the intensity of incident IR. The trend of responsivity and detectivity as a function of light intensities is reported in Figure 4d where a saturation of both parameters is evident at high intensity because the carrier recombination rate is reduced at high light intensity [92].

Recently, CNT/Si photodetectors in two electrode configurations for photovoltaic and photoconductive operations were studied under nanosecond light pulse [93]. A linear dependence of the photocurrent as a function of the light pulse energy was observed in photovoltaic mode with rise time of 20 ns, while when the device operates in photoconductive mode, the maximum photocurrent increases up to 30 times with a gain in the number of photogenerated charges till 200% and a reduction in the time response below 10 ns. At carbon nanotube/Si interface a Schottky junctions is formed, leading to a fast response, as pointed out by the current-voltage characteristics measured as a function of the temperature [93]. These findings open the possibility of CNT/Si photodetectors used as avalanche photomultipliers.

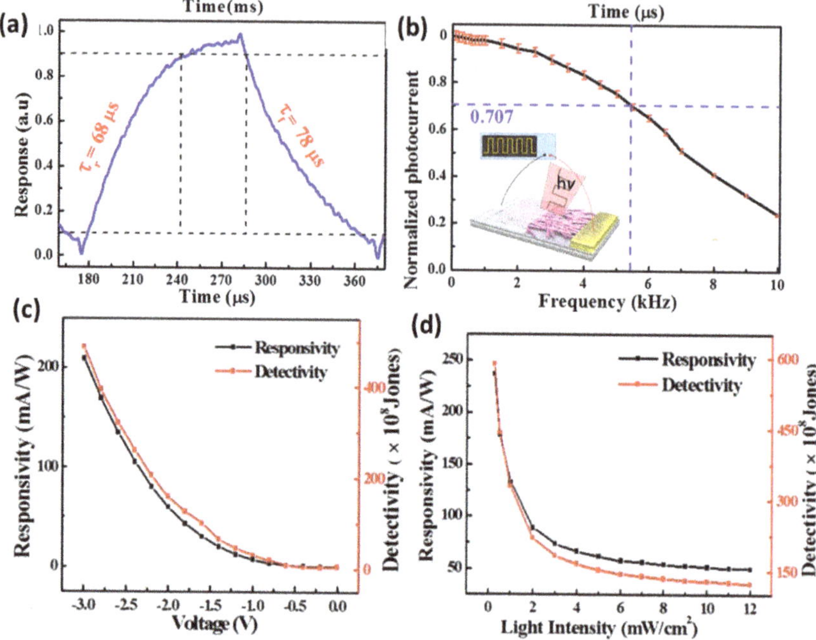

Figure 4. Photoresponse of the PD. (**a**) Single normalized cycle measured at 5 kHz for evaluating both rise (τ_r) and fall time (τ_f). (**b**) Normalized photocurrent versus switching frequency. (**c**) Responsivity and detectivity of the PD as a function of bias voltage. (**d**) Responsivity and detectivity of the PD as a function of light intensity. (Adapted from [80]).

3.3.2. Graphene

Since its discovery in 2004, graphene, a single layer of carbon atoms arranged in a closely packed two-dimensional honeycomb lattice, has attracted a lot of research interests due to its intriguing physical properties such as ultrahigh mobility (200,000 cm^{-2}/V·s at room temperature), high electrical conductivity (10^8 S/m) and an exceptionally large thermal conductivity up to 5300 W·m^{-1}·K^{-1} [94]. Moreover, graphene has would be involved in fabricating thinner and faster response speed optoelectronic devices [95,96]. Conversely, even if graphene can act as an absorber in ultrafast and ultra-broadband detectors, graphene-based photodetectors show relatively low responsivity, since a single sheet of carbon atoms has a relatively low absorbance (only 2.3%) from the ultraviolet (UV) to near infrared (NIR) region and short light-matter interaction length. Moreover, in pure graphene, due to its gapless nature, lifetime of excitons is extremely short giving rise to fast carrier recombination, which limits the efficient generation of photocurrent or photovoltage [97,98].

To overcome this problem, Schottky junction based photodiodes have been proposed in several combination of nanostructures such as II-VI (ZnO [99,100], ZnTe [101]), IV (Si [102,103], Ge [104], GeTe [105]), and III-VI (GaN [106], InAs [107], Al$_2$O$_3$ [108]), leading to improvements of both sensitivity and response speed. A very fascinating feature of the Schottky junction is that, due to its photovoltaic characteristics, most of the semiconductor-graphene devices are able to detect light irradiation without power supply [80]. An interesting and complete review of PDs based on graphene/semiconductor hybrid heterostructures, comprising device physics, design, performance, and process technologies for the optimization of PDs was recently published by Shin and Choi [109]. Considering that generally an all-Si approach is desired to take advantage by the existing CMOS technologies, a detailed review on the emerging field of the NIR internal photoemission effect-based graphene/Si PDs is presented by Casalino [110].

Regarding silicon-graphene heterostructures, in 2018, Periyanagounder et al. [111] realized and characterized a graphene/silicon (Gr/Si) (2D/3D) van der Waals heterostructure for high-performance PDs. Here, graphene works as an efficient carrier collector and Si as a photon absorption layer, allowing to obtain a barrier height of 0.76 eV and good performance as a self-powered detector under the mechanism of photovoltaic effect, working at 532 nm with zero bias. The authors measured a responsivity up to 510 mA W^{-1}, a photo switching ratio of 10^5 and a response time of 130 µs. These good results have been ascribed to the Schottky barrier which extends the lifetime of photo-excited carriers leading to a fast separation and transport of photoexcited carriers [111].

More recently, a compact PD operating as a metal–graphene–metal photoconductive detector has been co-integrated with a Si photonic waveguide, demonstrating extremely efficient and high-speed plasmonically enhanced waveguide-integrated photodetector [112]. A Si waveguide allows the light enters and evanescently coupled into the graphene-based PD, made with a 6 µm long layer of graphene. Then, the in-plane electric field is enhanced by the integrated nanosized bowtie-shaped metallic structures, as reported in Figure 5a,b. Simulation of magnitude of x and y components of the electric field was performed, and results are shown in Figure 5c, where is clearly visible a strong plasmonic field enhancement along the edges and in the gap of the bowtie-shaped structures. The single-layer graphene device exhibits a high external responsivity of 0.5 A/W under a bias of −0.4 V as measured with an input power of 80 µW, while if double-layer graphene device is considered, a higher input power can be reached obtaining 0.4 A/W at a bias of −0.6 V for an input signal of 1.3 mW (see Figure 5d,e). Moreover, the device demonstrates a fast photoresponse up to at least 110 GHz [112].

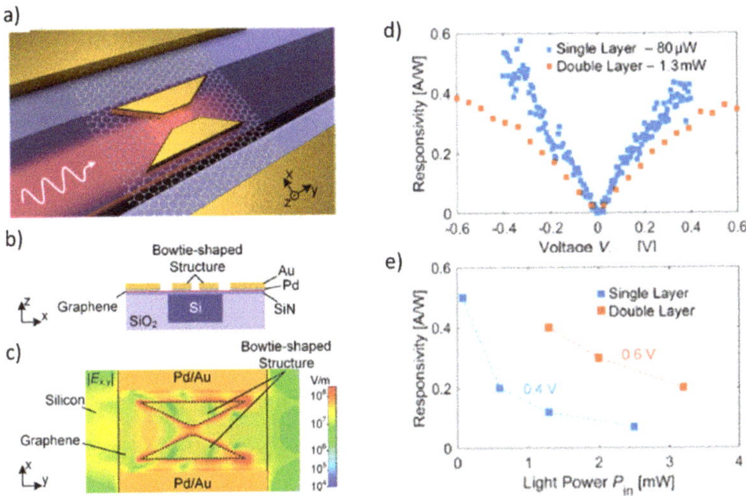

Figure 5. Plasmonically enhanced waveguide-integrated graphene PD. (**a**) 3D perspective view of the PD. (**b**) Cross sectional view of the graphene PD. (**c**) In-plane view of the magnitude of Ex and Ey electric fields, obtained by 3D full-wave finite-element method (FEM) simulations. (**d**) Responsivity of the PD as a function of bias voltage for a single-layer (blue scatters) and a double-layer graphene device (red scatters). (**e**) Responsivity of the PD as a function of light input power. (Adapted from [112]).

The results obtained so far, together with the growing study of high performance self-powered silicon-graphene heterostructures detectors paves the way for the technological implementation of Si-based monolithic optoelectronic devices for highest speed communication applications.

4. Colloidal Semiconductor

As mentioned before, materials nanostructuring offers the possibility to enhance material properties and devices performances. In the case of PDs, bottom up and top down fabrication of nanocrystals (NCs), nanoparticles (NPs), nanowires (NWs) etc. have been largely experienced to improve resistivity, work function (WF), bandgap or dielectric constant (ε). At the same time, dopant percentage, nanocomposition and several kind of deposition methods have been exploited to realize PD with low dark current, high EQE and wide band absorption.

4.1. Metal Oxide

Colloidal metal oxides have attracted much attention as based semiconductor material, because of their high stability [24,113,114]. Moreover, colloidal metal oxides showed adjustable electrical properties as resistivity, work function (WF) and dielectric constant (ε), which can be tuned by controlling vacancy density, doping material or other controllable parameters [115]. Finally, their showed tunable optical properties (e.g., photoluminescence, absorption, and transparency) tuning the band gap by changing alloys and engineering of components [116,117].

In PDs, colloidal metal oxides are largely employed because of their wide band gap, large exciton binding energy, combined with the simplicity of fabrication on a large-scale, with low cost, and tunable physical and chemical properties [118–121].

Also in this case, PDs can be divided into planar PDs and vertical PDs, according to device design [122,123]. PDs in planar configuration consist of an electrode and one only active layers (MOs), as shown in Figure 6a. Under no illumination, planar PDs have small Schottky barriers at the contacts, while when they are illuminated, photons are absorbed and electron–hole pairs are created, resulting in a photocurrent, by applying a voltage. The applied voltage reduce the carrier transit time and the photocurrent increases (Figure 6a) [124]. In vertical PDs, a n-doped semiconductor layer and a p-doped semiconductor layer are added, as shown in Figure 6b. Under illumination, the doped layers transport electrons and holes, decreasing the carrier transit time and, thus, increasing the efficiency of the device (Figure 6b) [125].

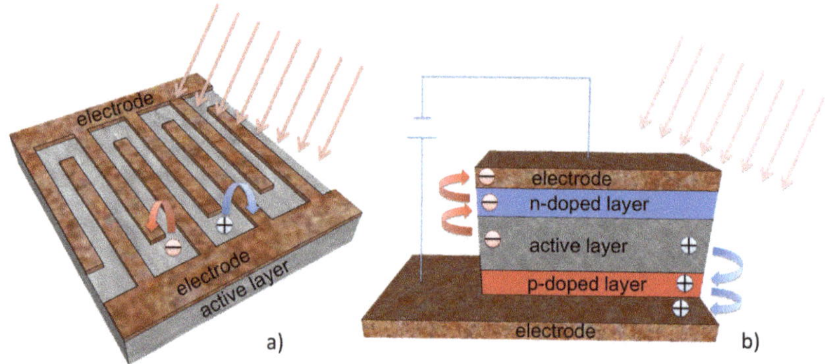

Figure 6. (**a**) Planar configuration and its basic working operation; (**b**) vertical configuration and its basic working operation.

Focusing on the active layer in planar PDs, in recent years, a lot of efforts have been employed to syntetize colloidal wide band gap MOs such as TiO_2 [126] and ZnO [127], used in UV PDs [128,129]. Nowadays, colloidal nanowires (NW) and nanobelts (NB) offer excellent solutions for optoelectronic applications and in particular for PD.

In [130], the authors synthesize SnO_2 NWs with a quite uniform diameter of about 26 nm, assemble a film by dip coating and fabricate a high performance PD device. In fact, it has excellent selectivity and light stability. Moreover, in [130] the EQE has been calculated with the follow equation:

$$EQE = \frac{t_{life}}{t_{tran}} = \frac{hc}{e\lambda} \cdot \frac{\Delta I}{PS} \tag{5}$$

where t_{life} and t_{tran} are the lifetime of carriers and the transit time between electrodes, respectively, h is he Planck's constant, c is the velocity of light, e is the electronic charge, λ is the incident wavelength, ΔI is the difference between the current and dark current, P is the light power density and S is the irradiated area of a single nanowire. For the present device, irradiated by 320 nm light at 0.91 mW cm^{-2}, EQE has been calculated and a value of 1.32×10^7 was obtained, four orders of magnitude larger than that of standard SnO_2 photodetectors. This detector is proposed not only for optoelectronic applications, but also in the visible and UV spectrum.

The PD assembled in a film of $In_2Ge_2O_7$ nanobelt with high quality monocrystals are expected to find wide optoelectronic applications in switches, optical sensors and generally in communication systems. In [131], the developed detectors showed high optical and electrical performance, i.e., EQE of 2.0×10^8%, decay times of ~3 ms and responsivity of 3.9×10^5 A/W, in addiction to high stability, reproducibility, sensitivity and selectivity.

PD based on NB of Nb_2O_5 showed high photosensitivity, selectivity to light and excellent photocurrent stability for >2500 s. The responsivity and EQE are determined to reach 15.2 A/W and 6070% respectively, demonstrating the potential of NB Nb_2O_5 in next-generation photosensors and PDs into the UV spectrum [132].

Zn_2SnO_4 (ZTO) has aroused great interest as a base material for PDs for its photodetection performance. In particular, ZTO is used as a PD in association with various polymers that are used to decorate the surface of ZTO, improving its ZTO properties.

In [133], it has been shown that by combining ZTO's NWs with polydimethylsiloxane (PDMS) it is possible to exploit the transparency and flexibility of PDMS with its NWs photodetection functionality. Furthermore, the photoactive materials are chemically bound in the PDMS matrix and less exposed to air and therefore to oxidation. Compared to the NWs completely immersed in the polymer matrix, the decorated NWs have a significant improvement in the PD in terms of speed of response (less than 0.8) with a recovery time of around 3 s.

Another type of decoration of NWs has been presented by Li et al. in [134]. The authors manufactured a PD with heterojunction structure with high performance in the UV spectrum, in which the active material is made up of ZTO NWs decorated with quantum dots (QDs) in ZnO. the NWs decorated with ZnO are summarized in two steps. In Figure 7, a sketch of carriers separation in ZnO QD decorated ZTO NW PD on lighting is shown.

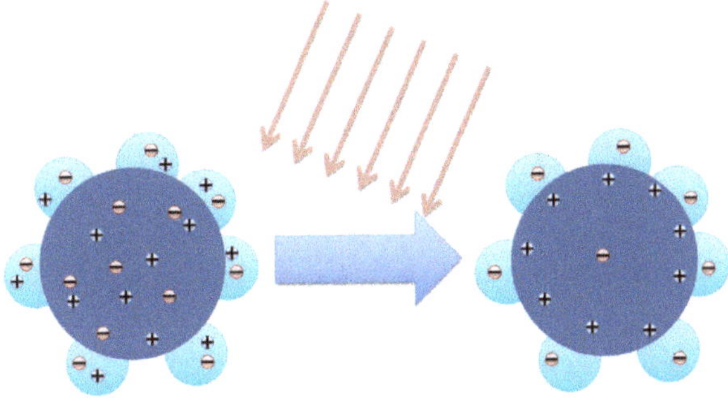

Figure 7. Sketch of carrier separation in the cross section of ZTO nanowires (NWs) decorated with ZnO quantum dots (QDs) upon illumination.

Compared to simple ZTO NWs, decorated QD NWs has 10 times higher photocurrent and response speed, with a light-dark current ratio up to 6.8×10^4 and a photoconductive gain up to 1.1×10^7, in addition to excellent stability.

Decoration of ZnO nanoparticles (NPs) has been also used to broader the absorption band and improve the photosensitivity. As example, in ZnO NPs the excess of Zn^{2+} ions toghedar with oxygen lack cause a great green photoluminescence [135,136], but coating the ZnO NPs with polyvinyl-alcohol, as done in [137], the ratio of photocurrent on dark current ranges from 3.8×10^6 to 1.34×10^8. In [138] simply changing the solution method of ZnO NPs into a coating spray, the authors shows that it is possible to tune the percentage of vacuum and zinc interstitial spaces. In this way, a significant broadening in visible light photo-detection is reached. Finally, in [139], Gogurla et al. synthesized gold decorated ZnO NPs colloidal spin coating solution. The fabricated film shows strong plasmon assisted SSA and scattering, leading to an increase of photoresponse by 80 times over the control simple ZnO NPs.

In the end of this paragraph, we would underline how colloidal NPs can be also easily integrated in a Si based heterojunction by drop casting; as done in [140], where authors synthesized In_2O_3 NPs and casted them on an n-type single-crystal Si wafer, showed an high photoresponse in visible and near-infrared regions. Infact in [140], in order to fabricate a heterojunction photodetector, thin films In_2O_3 NPs is deposited on a (100)-oriented 10 Ω cm n-type silicon by drop-casting and dried at 80 °C for 15 min under vacuum with multiple layers deposition and condensation steps. The final thickness of the In_2O_3 layer was 150 nm. Aluminum electrodes have been fabricated by means of standard lithographic process to make the ohmic contacts on the In_2O_3 nanoparticles layer and the back side of the Si wafer

4.2. HgTe and HgSe Nanocristals

In the infrared region, Hg nanocomposites have been widely explored to produce active materials in both planar and vertical PD. In particular, HgTe colloidal nanocrystals have been investigated in PD at telecom wavelength [141–143] and for solar applications [144,145] for their photoconduction properties [146].

In the NIR, typical applications are in communications technologies, biological imaging, and night imaging. Among possible colloidal nanomaterials to be used in NIR, HgTe NCs have reached enough maturity. In those materials, interband transitions assure IR absorption, see Figure 8 from [23].

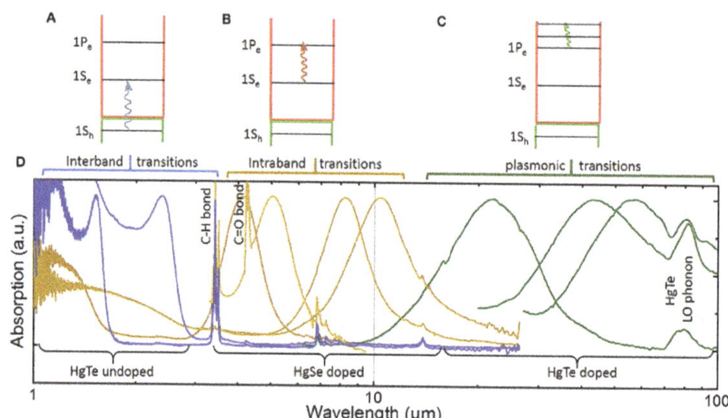

Figure 8. (**A–C**) Scheme for interband, intraband, and plasmonic transitions, respectively, in HgX NCs. (**D**) Absorption spectra HgX NCs (reprinted from [23]).

For wavelengths up to 1.7 µm, performances of InGaAs based PD are again higher. However, Hg based nanocomposites are promising as extending the range of wavelengths and reducing cost [147]. In fact, if in PD based on InGaAs is possible to achieve extremely low dark current at room temperature (<20 fA), in colloidal NCs based devices currently reported, such dark current are not so low. However, colloidal NCs based PDs have typically faster detection times, then they are usually employed to flame detection [148], night imaging [149] and biological imaging [150] for which detection is mandatory sub-ms sampling times.

In this case, the using of decorated NPs, i.e., coupling the absorption of the HgX NCs with plasmonic structures, show improvements in detectivity, as done in [151–153]. In particular, coupling HgTe with gold nanorods, the absorption is increased respect to thin layers of colloidal NCs, and a three time detectivity is reported respect the undecorated ones.

However, the use of colloidal NCs has some challenges that need to be addressed to bring the IR NCs based PDs to industrial production. Main challenges are about materials properties, as well as the toxicity and the trial of large synthesis (>10 g). Moreover, the not well known band structure of new NCs implies difficulties to design device with optimized ohmic contacts and generally band alignments; the HgX NCs experience rapid oxidation in air, then either the NCs surface have to be chemically stabilized in air, or the material have to be processed in glow box and then encapsulated into a protective layer.

5. Conclusions and Outlook

Photodetectors can be used in a wide range of applications, such as medical imaging, nuclear safety guards, aerospace, high energy physics experiments, photonics, and astrophysics. For this reason, there is a great interest in developing high performance, cheap and compact light sensor devices.

In this article, we firstly reviewed recent research progress in the field of photodetectors based on Group IV semiconductors. Historically, Group IV semiconductors are the most used materials for electronics devices; this is especially true for silicon, which is considered the main material for high integration CMOS technology. For this reason, we have investigated the developments of photodetectors based on these semiconductors and presented in recent years. Several narrow band-gap elementary semiconductors (e.g., Si, Ge, graphene, and CNTs) have been considered and different solutions proposed in the recent literature are reviewed in order to obtain photodetectors with good performance. Overall, MBA-based waveguide photodetectors are showing better performance than devices based on other absorption mechanisms. The recent reduction in the size ability of NIR all-Si photodetector production has open the road to new structures such as SSA and MBA-based ring resonator PDs or TPA-based PhC nanocavity PDs.

Then, an overview on colloidal semiconductor based photodetectors has been also presented; as these materials represents the frontier in future research, promising new opportunities for high performance and high-density chip-scale photonic integration from UV to infrared spectral regions.

The two semiconductor groups taken into consideration represent one the story that can still evolve and bring new improvements and the other the central topic on which to invest in the future, respectively. However, additional studies are required to further improve the device performance.

We strongly believe that high performance photodetectors based on group IV or colloidal semiconductors monolithically integrated on Si showing better functionality and high integration density, may be proven in the next future.

Author Contributions: P.D. and M.A.F.: writing—original draft preparation, review and editing. Both authors have read and agreed to the published version of the manuscript.

Funding: This research received no external funding.

Conflicts of Interest: The authors declare no conflict of interest.

References

1. Jamroz, W.; Kruzelecky, R.; Haddad, E. *Applied Microphotonics*; CRC Press: Boca Raton, FL, USA, 2006; p. 1. ISBN 9780849340260.
2. Soref, R.A. Microphotonics in-and-on silicon (state of the art and perspectives). In *Silicon-Based Microphotonics: From Basics to Applications*; Bisi, O., Campisano, L., Pavesi, L., Priolo, F., Eds.; IOS Press: Amsterdam, The Netherlands, 1999; pp. 1–20.
3. Dardano, P.; Moretti, L.; Mocella, V.; Sirleto, L.; Rendina, I. Investigation of a tunable T-shaped waveguide based on a silicon 2D photonic crystal. *J. Opt. A Pure Appl. Opt.* **2006**, *8*, S554. [CrossRef]
4. Mocella, V.; Cabrini, S.; Chang, A.S.P.; Dardano, P.; Moretti, L.; Rendina, I.; Olynick, D.; Harteneck, B.; Dhuey, S. Self-collimation of light over millimeter-scale distance in a quasi-zero-average-index metamaterial. *Phys. Rev. Lett.* **2009**, *102*, 133902. [CrossRef] [PubMed]
5. Dardano, P.; Gagliardi, M.; Rendina, I.; Cabrini, S.; Mocella, V. Ellipsometric determination of permittivity in a negative index photonic crystal metamaterial. *Light Sci. Appl.* **2012**, *1*, e42. [CrossRef]
6. Dardano, P.; Borrelli, M.; Musto, M.; Rotondo, G.; Iodice, M. Computational analysis of cooling dynamics in photonic-crystal-based thermal switches. *J. Eur. Opt. Soc.-Rapid Publ.* **2016**, *12*, 1–7. [CrossRef]
7. Dardano, P.; Caliò, A.; Politi, J.; Rea, I.; Rendina, I.; De Stefano, L. Optically monitored drug delivery patch based on porous silicon and polymer microneedles. *Biomed. Opt. Express* **2016**, *7*, 1645–1655. [CrossRef]
8. Dardano, P.; Caliò, A.; Di Palma, V.; Bevilacqua, M.F.; Di Matteo, A.; De Stefano, L. A photolithographic approach to polymeric microneedles array fabrication. *Materials* **2015**, *8*, 8661–8673. [CrossRef]
9. Politi, J.; Dardano, P.; Caliò, A.; Iodice, M.; Rea, I.; De Stefano, L. Reversible sensing of heavy metal ions using lysine modified oligopeptides on porous silicon and gold. *Sens. Actuators B Chem.* **2017**, *244*, 142–150. [CrossRef]
10. Sirleto, L.; Ferrara, M.A.; Nikitin, T.; Novikov, S.; Khriachtchev, L. Giant Raman gain in silicon nanocrystals. *Nat. Commun.* **2012**, *3*, 1220. [CrossRef]
11. Ferrara, M.A.; Rendina, I.; Basu, S.N.; Dal Negro, L.; Sirleto, L. Raman amplifier based on amorphous silicon nanoparticles. *Int. J. Photoenergy* **2012**, *2012*, 254946. [CrossRef]
12. Ferrara, M.A.; Sirleto, L. Integrated Raman laser: A review of the last two decades. *Micromachines* **2020**, *11*, 330. [CrossRef]
13. Hui, R. Photodetectors. In *Introduction to Fiber-Optic Communications*; Academic Press: Cambridge, MA, USA, 2020; pp. 125–154. [CrossRef]
14. Xing, G.; Mathews, N.; Sun, S.; Lim, S.S.; Lam, Y.M.; Grätzel, M.; Mhaisalkar, S.; Sum, T.C. Long-Range Balanced Electron- and Hole-Transport Lengths in Organic-Inorganic $CH_3NH_3PbI_3$. *Science* **2013**, *342*, 344–347. [CrossRef] [PubMed]
15. Rogalski, A.; Antoszewski, J.; Faraone, L. Third-generation infrared photodetector arrays. *J. Appl. Phys.* **2009**, *105*, 091101. [CrossRef]
16. Schneider, H.; Liu, H.C. *Quantum Well Infrared Photodetectors, Physics and Applications*; Springer: Heidelberg, Germany, 2006.
17. Berryman, K.W.; Lyon, S.A.; Segev, M. Mid-infrared photoconductivity in InAs quantum dots. *Appl. Phys. Lett.* **1997**, *70*, 1861–1863. [CrossRef]
18. Pan, D.; Towe, E.; Kennerly, S. Normal-incidence intersubband (In, Ga)As/GaAs quantum dot infrared photodetectors. *Appl. Phys. Lett.* **1998**, *73*, 1937–1939. [CrossRef]
19. Sauvage, S.; Boucaud, P.; Brunhes, T. Midinfrared absorption and photocurrent spectroscopy of InAs/GaAs self-assembled quantum dots. *Appl. Phys. Lett.* **2001**, *78*, 2327–2329. [CrossRef]
20. Chakrabarti, S.; Stiff-Roberts, A.D.; Su, X.H.; Bhattacharya, P.; Ariyawansa, G.; Perera, A.G.U. High-performance mid-infrared quantum dot infrared photodetectors. *J. Phys. D Appl. Phys.* **2005**, *38*, 2135–2141. [CrossRef]
21. Lim, H.; Tsao, S.; Zhang, W.; Razeghi, M. High-performance quantumdot infrared photodetectors grown on InP substrate operating at room temperature. *Appl. Phys. Lett.* **2007**, *90*, 131112. [CrossRef]
22. Martyniuk, P.; Rogalski, A. Quantum-dot infrared photodetectors: Status and outlook. *Prog. Quantum Electron.* **2008**, *32*, 89–120. [CrossRef]
23. Livache, C.; Martinez, B.; Goubet, N.; Ramade, J.; Lhuillier, E. Road Map for Nanocrystal Based Infrared Photodetectors. *Front. Chem.* **2018**, *6*, 575. [CrossRef]

24. Song, J. Colloidal metal oxides in electronics and optoelectronics. In *Colloidal Metal Oxide Nanoparticles*; Sabu, T., Anu, T.S., Prajitha, V., Eds.; Elsevier: Amsterdam, The Netherlands, 2020; pp. 203–246.

25. Yotter, R.A.; Wilson, D.M. A review of photodetectors for sensing light-emitting reporters in biological systems. *IEEE Sens. J.* **2003**, *3*, 288–303. [CrossRef]

26. Ren, A.; Yuan, L.; Xu, H.; Wu, J.; Wang, Z. Recent progress of III–V quantum dot infrared photodetectors on silicon. *J. Mater. Chem. C* **2019**, *7*, 14441–14453. [CrossRef]

27. Stöckmann, F. Photodetectors, their performance and their limitations. *Appl. Phys.* **1975**, *7*, 1–5. [CrossRef]

28. Grinberg, A.A.; Luryi, S. Theory of the photon-drag effect in a two-dimensional electron gas. *Phys. Rev. B* **1988**, *38*, 87–96. [CrossRef] [PubMed]

29. Bishop, P.; Gibson, A.; Kimmitt, M. The performance of photon-drag detectors at high laser intensities. *IEEE J. Quantum Electron.* **1973**, *9*, 1007–1011. [CrossRef]

30. Herrscher, M.; Grundmann, M.; Dröge, E.; Bottcher, E.H.; Bimberg, D. Epitaxial liftoff InGaAs/InP MSM photodetectors on Si. *Electron. Lett.* **1995**, *31*, 1383–1384. [CrossRef]

31. Walden, R.H. A review of recent progress in InP-based optoelectronic integrated circuit receiver front-ends. *Int. J. High Speed Electron. Syst.* **1998**, *9*, 631–642. [CrossRef]

32. Masini, G.; Cencelli, V.; Colace, L.; De Notaristefani, F.; Assanto, G. Monolithic integration of near-infrared Ge photodetectors with Si complementary metal-oxide-semiconductor readout electronics. *Appl. Phys. Lett.* **2002**, *80*, 3268–3270. [CrossRef]

33. Chui, C.O.; Okyay, A.K.; Saraswat, K.C. Effective dark current suppression with asymmetric MSM photodetectors in Group IV semiconductors. *IEEE Photonics Technol. Lett.* **2003**, *15*, 1585–1587. [CrossRef]

34. Wohlmuth, W.A.; Arafa, M.; Mahajan, A.; Fay, P.; Adesida, I. InGaAs metal-semiconductor-metal photodetectors with engineered Schottky barrier heights. *Appl. Phys. Lett.* **1996**, *69*, 3578–3580. [CrossRef]

35. Soole, J.B.D.; Schumacher, H. InGaAs metal-semiconductor-metal photodetectors for long wavelength optical communications. *IEEE J. Quantum Electron.* **1991**, *27*, 737–752. [CrossRef]

36. Koester, S.J.; Schaub, J.D.; Dehlinger, G.; Chu, J.O. Germanium-on-SOI infrared detectors for integrated photonic applications. *IEEE J. Sel. Top. Quantum Electron.* **2006**, *12*, 1489–1502. [CrossRef]

37. Harame, D.L.; Koester, S.J.; Freeman, G.; Cottrel, P.; Rim, K.; Dehlinger, G.; Ahlgren, D.; Dunn, J.S.; Greenberg, D.; Joseph, A.; et al. The revolution in SiGe: Impact on device electronics. *Appl. Surf. Sci.* **2004**, *224*, 9–17. [CrossRef]

38. Wang, J.; Lee, S. Ge-Photodetectors for Si-Based Optoelectronic Integration. *Sensors* **2011**, *11*, 696–718. [CrossRef] [PubMed]

39. Zhu, S.; Chu, H.S.; Lo, G.Q.; Bai, P.; Kwong, D.L. Waveguide-integrated near-infrared detector with self-assembled metal silicide nanoparticles embedded in a silicon p-n junction. *Appl. Phys. Lett.* **2012**, *100*, 061109. [CrossRef]

40. Berini, P.; Olivieri, S.; Chen, C. Thin Au surface plasmon waveguide Schottky detectors on p-Si. *Nanotechnology* **2012**, *23*, 444011. [CrossRef]

41. Akbari, A.; Tait, R.N.; Berini, P. Surface plasmon waveguide Schottky detector. *Opt. Express* **2010**, *18*, 8505–8514. [CrossRef]

42. Sobhani, A.; Knight, M.W.; Wang, Y.; Zheng, B.; King, N.S.; Brown, L.V.; Fang, Z.; Nordlander, P.; Halas, N.J. Narrowband photodetection in the near-infrared with a plasmon-induced hot electron device. *Nat. Commun.* **2013**, *4*, 1643. [CrossRef]

43. Amirmazlaghani, M.; Raissi, F.; Habibpour, O.; Vukusic, J.; Stake, J. Graphene-Si Schottky IR detector. *IEEE J. Quant. Electron.* **2013**, *49*, 589–594. [CrossRef]

44. Casalino, M.; Sirleto, L.; Iodice, M.; Coppola, G. Silicon photodetectors: The challenge of detecting near-infrared light. In *Photodetectors*; Gateva, S., Ed.; InTech: London, UK, 2012; pp. 51–76. [CrossRef]

45. Casalino, M. Recent Advances in Silicon Photodetectors Based on the Internal Photoemission Effect. In *New Research on Silicon-Structure, Properties, Technology*; Talanin, V.I., Ed.; InTech: London, UK, 2017; pp. 245–266. [CrossRef]

46. Kimata, M.; Ueno, M.; Yagi, H.; Shiraishi, T.; Kawai, M.; Endo, K.; Kosasayama, Y.; Sone, T.; Ozeki, T.; Tsubouchi, N. PtSi Schottky-barrier infrared focal plane arrays. *Opto-Electron. Rev.* **1998**, *6*, 1–10.

47. Sakib, M.; Sun, J.; Kumar, R.; Driscoll, J.; Yeung, K.; Rong, H. Demonstration of a 50 Gb/s all-silicon waveguide photodetector for photonic integration. In Proceedings of the Conference on Lasers and Electro-Optics, OSA Technical Digest (online) 2018, San Jose, CA, USA, 13–18 May 2018; p. JTh5A.7. [CrossRef]

48. Fang, X.S.; Bando, Y.; Liao, M.Y.; Gautam, U.K.; Zhi, C.Y.; Dierre, B.; Liu, B.D.; Zhai, T.Y.; Sekiguchi, T.; Koide, Y.; et al. Single-Crystalline ZnS Nanobelts as Ultraviolet-Light Sensors. *Adv. Mater.* **2009**, *21*, 2034–2039. [CrossRef]

49. Huang, X.; Wang, M.; Willinger, M.G.; Shao, L.D.; Su, D.S.; Meng, X.M. Assembly of three-dimensional hetero-epitaxial ZnO/ZnS core/shell nanorod and single crystalline hollow ZnS nanotube arrays. *ACS Nano* **2012**, *6*, 7333–7339. [CrossRef] [PubMed]

50. Tian, W.; Liu, D.; Cao, F.; Li, L. Hybrid Nanostructures for Photodetectors. *Adv. Opt. Mater.* **2017**, *5*, 1600468. [CrossRef]

51. Sun, J.; Han, M.; Gu, Y.; Yang, Z.-X.; Zeng, H. Recent Advances in Group III–V Nanowire Infrared Detectors. *Adv. Opt. Mater.* **2018**, *6*, 1800256. [CrossRef]

52. Yang, C.; Barrelet, C.J.; Capasso, F.; Lieber, C.M. Hybrid Single-Nanowire Photonic Crystal and Microresonator Structures. *Nano Lett.* **2006**, *6*, 2929. [CrossRef] [PubMed]

53. Liang, F.-X.; Wang, J.-Z.; Li, Z.-P.; Luo, L.-B. Near-Infrared-Light Photodetectors Based on One-Dimensional Inorganic Semiconductor Nanostructures. *Adv. Optical Mater.* **2017**, *5*, 1700081. [CrossRef]

54. Wu, C.Y.; Pan, Z.Q.; Wang, Y.Y.; Ge, C.W.; Yu, Y.Q.; Xu, J.Y.; Wang, L.; Luo, L.B. Core–shell silicon nanowire array–Cu nanofilm Schottky junction for a sensitive self-powered near-infrared photodetector. *J. Mater. Chem. C* **2016**, *4*, 10804–10811. [CrossRef]

55. Mulazimoglu, E.; Coskun, S.; Gunoven, M.; Butun, B.; Ozbay, E.; Turan, R.; Unalan, H.E. Silicon nanowire network metal-semiconductor-metal photodetectors. *Appl. Phys. Lett.* **2013**, *103*, 083114. [CrossRef]

56. Das, K.; Mukherjee, S.; Manna, S.; Raychaudhuri, A.K. Single Si nanowire (diameter ≤ 100 nm) based polarization sensitive near-infrared photodetector with ultra-high responsivity. *Nanoscale* **2014**, *6*, 11232–11239. [CrossRef]

57. Cao, Y.; Zhu, J.; Xu, J.; He, J.; Sun, J.L.; Wang, Y.; Zhao, Z. Ultra-Broadband Photodetector for the Visible to Terahertz Range by Self-Assembling Reduced Graphene Oxide-Silicon Nanowire Array Heterojunctions. *Small* **2014**, *10*, 2345–2351. [CrossRef]

58. Xie, C.; Nie, B.; Zeng, L.H.; Liang, F.X.; Wang, M.Z.; Luo, L.B.; Feng, M.; Yu, Y.Q.; Wu, C.Y.; Wu, Y.C.; et al. Core-shell heterojunction of silicon nanowire arrays and carbon quantum dots for photovoltaic devices and self-driven photodetectors. *ACS Nano* **2014**, *8*, 4015–4022. [CrossRef]

59. Yatsui, T.; Okada, S.; Takemori, T.; Sato, T.; Saichi, K.; Ogamoto, T.; Chiashi, S.; Maruyama, S.; Noda, M.; Yabana, K.; et al. Enhanced photo-sensitivity in a Si photodetector using a near-field assisted excitation. *Commun. Phys.* **2019**, *2*, 62. [CrossRef]

60. Huang, S.; Wu, Q.; Jia, Z.; Jin, X.; Fu, X.; Huang, H.; Zhang, X.; Yao, J.; Xu, J. Black Silicon Photodetector with Excellent Comprehensive Properties by Rapid Thermal Annealing and Hydrogenated Surface Passivation. *Adv. Opt. Mater.* **2020**, *8*, 1901808. [CrossRef]

61. Benedikovic, D.; Virot, L.; Aubin, G.; Hartmann, J.-M.; Amar, F.; Szelag, B.; Le Roux, X.; Alonso-Ramos, C.; Crozat, P.; Cassan, É.; et al. Comprehensive Study on Chip-Integrated Germanium Pin Photodetectors for Energy-Efficient Silicon Interconnects. *IEEE J. Quantum Electron.* **2020**, *56*, 8400409. [CrossRef]

62. Samavedam, S.B.; Currie, M.T.; Langdo, T.A.; Fitzgerald, E.A. High-quality germanium photodiodes integrated on silicon substrates using optimized relaxed graded buffers. *Appl. Phys. Lett.* **1998**, *73*, 2125–2127. [CrossRef]

63. Zhang, Y.; Yang, S.; Yang, Y.; Gould, M.; Ophir, N.; Eu-Jin Lim, A.; Lo, G.-Q.; Magill, P.; Bergman, K.; Baehr-Jones, T.; et al. A high-responsivity photodetector absent metalgermanium direct contact. *Opt. Exp.* **2014**, *22*, 11367–11375. [CrossRef]

64. Lischke, S.; Knoll, D.; Mai, C.; Zimmermann, L.; Peczek, A.; Kroh, M.; Trusch, A.; Krune, E.; Voigt, K.; Mai, A. High bandwidth, high responsivity waveguidecoupled germanium p-i-n photodiode. *Opt. Exp.* **2015**, *23*, 27213–27220. [CrossRef]

65. Virot, L.; Benedikovic, D.; Szelag, B.; Alonso-Ramos, C.; Karakus, B.; Hartmann, J.-M.; Le Roux, X.; Crozat, P.; Cassan, E.; Marris-Morini, D.; et al. Integrated waveguide PIN photodiodes exploiting lateral Si/Ge/Si heterojunction. *Opt. Exp.* **2017**, *25*, 19487–19496. [CrossRef]

66. Chen, H.; Verheyen, P.; De Heyn, P.; Lepage, G.; De Coster, J.; Balakrishnan, S.; Absil, P.; Yao, W.; Shen, L.; Roelkens, G.; et al. −1 V bias 67 GHz bandwidth Si-contacted germanium waveguide p-i-n photodetector for optical links at 56 Gbps and beyond. *Opt. Exp.* **2016**, *24*, 4622–4631. [CrossRef]

67. Benedikovic, D.; Virot, L.; Aubin, G.; Amar, F.; Szelag, B.; Karakus, B.; Hartmann, J.-M.; Alonso-Ramos, C.; Le Roux, X.; Crozat, P.; et al. 25 Gbps low-voltage hetero-structured silicongermanium waveguide pin photodetectors for monolithic on-chip nanophotonic architectures. *Photon. Res.* **2019**, *7*, 437–444. [CrossRef]

68. Tzu, T.; Sun, K.; Costanzo, R.; Ayoub, D.; Bowers, S.M.; Beling, A. Foundry-Enabled High-Power Photodetectors for Microwave Photonics. *IEEE J. Sel. Top. Quantum Electron.* **2019**, *25*, 3800111. [CrossRef]

69. Hu, W.; Cong, H.; Huang, W.; Huang, Y.; Chen, L.; Pan, A.; Xue, C. Germanium/perovskite heterostructure for high-performance and broadband photodetector from visible to infrared telecommunication band. *Light Sci. Appl.* **2019**, *8*, 106. [CrossRef]

70. Siontas, S.; Wang, H.; Li, D.; Zaslavsky, A.; Pacifici, D. Broadband visible-to-telecom wavelength germanium quantum dot photodetectors. *Appl. Phis. Lett.* **2018**, *113*, 181101. [CrossRef]

71. Kumar, S.; Chatterjee, A.; Selvaraja, S.K.; Avasthi, S. Two-Step Liquid Phase Crystallized Germanium-Based Photodetector for Near-Infrared Applications. *IEEE Sens. J.* **2020**, *20*, 4660–4666. [CrossRef]

72. Wang, Y. Photoconductivity of fullerene-doped polymers. *Nature* **1992**, *356*, 585–587. [CrossRef]

73. Baker, S.N.; Baker, G.A. Luminescent Carbon Nanodots: Emergent Nanolights. *Angew. Chem. Int. Ed.* **2010**, *49*, 6726–6744. [CrossRef]

74. Baughman, R.H.; Cui, C.; Zakhidov, A.A.; Iqbal, Z.; Barisci, J.N.; Spinks, G.M.; Wallace, G.G.; Mazzoldi, A.; De Rossi, D.; Rinzler, A.G.; et al. Carbon Nanotube Actuators. *Science* **1999**, *284*, 1340–1344. [CrossRef]

75. Zhang, G.L.; Zhou, R.L.; Zeng, X.C. Carbon nanotube and boron nitride nanotube hosted C-60-V nanopeapods. *J. Mater. Chem. C* **2013**, *1*, 4518–4526. [CrossRef]

76. Nair, R.R.; Blake, P.; Grigorenko, A.N.; Novoselov, K.S.; Booth, T.J.; Stauber, T.; Peres, N.M.R.; Geim, A.K. Fine structure constant defines visual transparency of graphene. *Science* **2008**, *320*, 1308. [CrossRef]

77. Ponomarenko, L.A.; Schedin, F.; Katsnelson, M.I.; Yang, R.; Hill, E.W.; Novoselov, K.S.; Geim, A.K. Chaotic Dirac billiard in graphene quantum dots. *Science* **2008**, *320*, 356–358. [CrossRef]

78. Huynh, W.U.; Dittmer, J.J.; Alivisatos, A.P. Hybrid nanorod-polymer solar cells. *Science* **2002**, *295*, 2425–2427. [CrossRef]

79. He, X.; Léonard, F.; Kono, J. Uncooled Carbon Nanotube Photodetectors. *Adv. Opt. Mater.* **2015**, *3*, 989–1011. [CrossRef]

80. Zhang, T.-F.; Li, Z.-P.; Wang, J.-Z.; Kong, W.-Y.; Wu, G.-A.; Zheng, Y.-Z.; Zhao, Y.-W.; Yao, E.-X.; Zhuang, N.-X.; Luo, L.-B. Broadband photodetector based on carbon nanotube thin film/single layer graphene Schottky junction. *Sci. Rep.* **2016**, *6*, 38569. [CrossRef]

81. Freitag, M.; Martin, Y.; Misewich, J.A.; Martel, R.; Avouris, P.H. Photoconductivity of single carbon nanotubes. *Nano Lett.* **2003**, *3*, 1067–1071. [CrossRef]

82. Lu, R.T.; Shi, J.J.; Baca, F.J.; Wu, J.Z. High performance multiwall carbon nanotube bolometers. *J. Appl. Phys.* **2010**, *108*, 084305. [CrossRef]

83. Chen, H.Z.; Xi, N.; Song, B.; Chen, L.; Zhao, J.; Lai, K.W.C.; Yang, R. Infrared Camera Using a Single Nano-Photodetector. *IEEE Sens. J.* **2013**, *13*, 949–958. [CrossRef]

84. Balasubramanian, K.; Burghard, M.; Kern, K.; Scolari, M.; Mews, A. Photocurrent Imaging of Charge Transport Barriers in Carbon Nanotube Devices. *Nano Lett.* **2005**, *5*, 507–510. [CrossRef]

85. Lee, J.U. Photovoltaic effect in ideal carbon nanotube diodes. *Appl. Phys. Lett.* **2005**, *87*, 073101. [CrossRef]

86. Avouris, P.; Freitag, M.; Perebeinos, V. Carbon-nanotube photonics and optoelectronics. *Nat. Photonics* **2008**, *2*, 341–350. [CrossRef]

87. Yang, L.; Wang, S.; Zeng, Q.; Zhang, Z.; Peng, L.M. Carbon nanotube photoelectronic and photovoltaic devices and their applications in infrared detection. *Small* **2013**, *9*, 1225–1236. [CrossRef]

88. Xie, Y.; Gong, M.; Shastry, T.A.; Lohrman, J.; Hersam, M.C.; Ren, S. Broad-Spectral-Response Nanocarbon Bulk-Heterojunction Excitonic Photodetectors. *Adv. Mater.* **2013**, *25*, 3433–3437. [CrossRef]

89. Bindl, D.J.; Wu, M.-Y.; Prehn, F.C.; Arnold, M.S. Efficiently Harvesting Excitons From Electronic Type-Controlled Semiconducting Carbon Nanotube Films. *Nano Lett.* **2011**, *11*, 455–460. [CrossRef] [PubMed]

90. Liang, L.; Ma, Z.; Wu, G.; Wei, N.; Huang, L.; Huang, H.; Liu, H.; Wang, S.; Peng, L.-M. Microcavity-Integrated Carbon Nanotube Photodetectors. *ACS Nano* **2016**, *10*, 6963–6971. [CrossRef] [PubMed]

91. Cheng, S.H.; Weng, T.-M.; Lu, M.-L.; Tan, W.-C.; Chen, J.-Y.; Chena, Y.-F. All carbon-based photodetectors: An eminent integration of graphite quantum dots and two dimensional grapheme. *Sci. Rep.* **2013**, *3*, 454. [CrossRef] [PubMed]

92. Zhang, H.B.; Zhang, X.J.; Liu, C.; Lee, S.T.; Jie, J.S. High-Responsivity, High-Detectivity, Ultrafast Topological Insulator Bi_2Se_3/Silicon Heterostructure Broadband Photodetectors. *ACS Nano* **2016**, *10*, 5113–5122. [CrossRef]

93. Salvato, M.; Scagliotti, M.; De Crescenzi, M.; Boscardin, M.; Attanasio, C.; Avallone, G.; Cirillo, C.; Prosposito, P.; De Matteins, F.; Messi, R.; et al. Time response in carbon nanotube/Si based photodetectors. *Sens. Actuators A Phys.* **2019**, *292*, 71–76. [CrossRef]

94. Balandin, A.A.; Ghosh, S.; Bao, W.; Calizo, I.; Teweldebrhan, D.; Miao, F.; Lau, C.N. Superior Thermal Conductivity of Single-Layer Graphene. *Nano Lett.* **2008**, *8*, 902–907. [CrossRef]

95. Koppens, F.H.L.; Mueller, T.; Ferrari, A.C.; Vitiello, M.S.; Polini, M. Photodetectors based on graphene, other two-dimensional materials and hybrid systems. *Nat. Nanotech.* **2014**, *9*, 780–793. [CrossRef]

96. Guo, N.; Hu, W.; Jiang, T.; Gong, F.; Luo, W.; Qiu, W.; Wang, P.; Liu, L.; Wu, S.; Liao, L.; et al. High-Quality Infrared Imaging with Graphene Photodetectors at Room Temperature. *Nanoscale* **2016**, *8*, 16065–16072. [CrossRef]

97. Xia, F.; Mueller, T.; Lin, Y.-M.; Valdes-Garcia, A.; Avouris, P. Ultrafast graphene photodetector. *Nat. Nanotechnol.* **2009**, *4*, 839–843. [CrossRef]

98. Bao, W.; Jing, L.; Velasco, J., Jr.; Lee, Y.; Liu, G.; Tran, D.; Standley, B.; Aykol, M.; Cronin, S.B.; Smirnov, D.; et al. Stacking-dependent band gap and quantum transport in trilayer graphene. *Nat. Phys.* **2011**, *7*, 948–952. [CrossRef]

99. Nie, B.; Hu, J.-G.; Luo, L.-B.; Xie, C.; Zeng, L.-H.; Lv, P.; Li, F.-Z.; Jie, J.-S.; Feng, M.; Wu, C.-Y.; et al. Monolayer Graphene Film on ZnO Nanorod Array for High-Performance Schottky Junction Ultraviolet Photodetectors. *Small* **2013**, *9*, 2872–2879. [CrossRef] [PubMed]

100. Dang, V.Q.; Trung, T.Q.; Kim, D.-I.; Duy, L.T.; Hwang, B.-U.; Lee, D.-W.; Kim, B.-Y.; Toan, L.D.; Lee, N.-E. Ultrahigh Responsivity in Graphene–ZnO Nanorod Hybrid UV Photodetector. *Small* **2015**, *11*, 3054–3065. [CrossRef] [PubMed]

101. Luo, L.B.; Zhang, S.H.; Lu, R.; Sun, W.; Fang, Q.L.; Wu, C.Y.; Hu, J.; Wang, L. P-type ZnTe:Ga nanowires: Controlled doping and optoelectronic device application. *RSC Adv.* **2015**, *5*, 13324–13330. [CrossRef]

102. Luo, L.B.; Zhang, S.-H.; Lu, R.; Sun, W.; Fang, Q.-L.; Wu, C.-Y.; Hua, J.-G.; Wang, L. Light trapping and surface plasmon enhanced high-performance NIR photodetector. *Sci. Rep.* **2014**, *4*, 3914. [CrossRef] [PubMed]

103. Goykhman, I.; Sassi, U.; Desiatov, B.; Mazurski, N.; Milana, S.; de Fazio, D.; Eiden, A.; Khurgin, J.; Shappir, J.; Levy, U.; et al. On-Chip Integrated, Silicon–Graphene PlasmonicSchottky Photodetector with High Responsivity and Avalanche Photogain. *Nano Lett.* **2015**, *16*, 3005–3013. [CrossRef]

104. Zeng, L.H.; Wang, M.-Z.; Hu, H.; Nie, B.; Yu, Y.-Q.; Wu, C.-Y.; Wang, L.; Hu, J.-G.; Xie, C.; Liang, F.-X.; et al. Monolayer graphene/germanium Schottky junction as high-performance self-driven infrared light photodetector. *ACS Appl. Mater. Interfaces* **2013**, *5*, 9362–9366. [CrossRef] [PubMed]

105. Petronijevic, E.; Leahu, G.; Di Meo, V.; Crescitelli, A.; Dardano, P.; Coppola, G.; Esposito, E.; Rendina, I.; Miritello, M.; Grimaldi, M.G.; et al. Near-infrared modulation by means of GeTe/SOI-based metamaterial. *Opt. Lett.* **2019**, *44.6*, 1508–1511. [CrossRef]

106. Kumar, A.; Kashild, R.; Ghosh, A.; Kumar, V.; Singh, R. Enhanced Thermionic Emission and Low 1/f Noise in Exfoliated Graphene/GaN Schottky Barrier Diode. *ACS Appl. Mater. Interfaces* **2016**, *8*, 8213–8223. [CrossRef]

107. Miao, J.S.; Hu, W.; Guo, N.; Lu, Z.; Liu, X.; Liao, L.; Chen, P.; Jiang, T.; Wu, S.; Ho, J.C.; et al. High-Responsivity Graphene/InAs Nanowire Heterojunction Near-Infrared Photodetectors with Distinct Photocurrent On/Off Ratio. *Small* **2015**, *11*, 936–942. [CrossRef]

108. Xu, J.; Liu, T.; Hu, H.; Zhai, Y.; Chen, K.; Chen, N.; Li, C.; Zhang, X. Design and optimization of tunneling photodetectors based on graphene/Al2O3/silicon heterostructures. *Nanophotonics* **2020**, *20190499*, 2192–8614. [CrossRef]

109. Shin, D.H.; Choi, S.-H. Graphene-Based Semiconductor Heterostructures for Photodetectors. *Micromachines* **2018**, *9*, 350. [CrossRef] [PubMed]

110. Casalino, M. Silicon Meets Graphene for a New Family of Near-Infrared Schottky Photodetectors. *Appl. Sci.* **2019**, *9*, 3677. [CrossRef]

111. Periyanagounder, D.; Gnanasekar, P.; Varadhan, P.; He, J.-H.; Kulandaivel, J. High performance, self-powered photodetectors based on a graphene/silicon Schottky junction diode. *J. Mater. Chem. C* **2018**, *6*, 9545–9551. [CrossRef]

112. Ma, P.; Salamin, Y.; Baeuerle, B.; Josten, A.; Heni, W.; Emboras, A.; Leuthold, J. Plasmonically Enhanced Graphene Photodetector Featuring 100 Gbit/s Data Reception, High Responsivity, and Compact Size. *ACS Photonics* **2019**, *6*, 154–161. [CrossRef]

113. Keller, A.A.; Wang, H.; Zhou, D.; Lenihan, H.S.; Cherr, G.; Cardinale, B.J.; Miller, R.; Ji, Z. Stability and aggregation of metal oxide nanoparticles in natural aqueous matrices. *Environ. Sci. Technol.* **2010**, *44*, 1962–1967. [CrossRef] [PubMed]

114. ELeite, R.; Maciel, A.P.; Weber, I.T.; Lisboa-Filho, P.N.; Longo, E.; Paiva-Santos, C.O.; Andrade, A.V.C.; Pakoscimas, C.A.; Maniette, Y.; Schreiner, H.W. Development of metal oxide nanoparticles with high stability against particle growth using a metastable solid solution. *Adv. Mater.* **2002**, *14*, 905–908.

115. Buonsanti, R.; Llordes, A.; Aloni, S.; Helms, B.A.; Milliron, D.J. Tunable infrared absorption and visible transparency of colloidal aluminum-doped zinc oxide nanocrystals. *Nano Lett.* **2011**, *11*, 4706–4710. [CrossRef]

116. Farvid, S.S.; Wang, T.; Radovanovic, P.V. Colloidal gallium indium oxide nanocrystals: A multifunctional light-emitting phosphor broadly tunable by alloy composition. *J. Am. Chem. Soc.* **2011**, *133*, 6711–6719. [CrossRef]

117. De Trizio, L.; Buonsanti, R.; Schimpf, A.M.; Llordes, A.; Gamelin, D.R.; Simonutti, R.; Milliron, D.J. Nb-doped colloidal TiO_2 nanocrystals with tunable infrared absorption. *Chem. Mater.* **2013**, *25*, 3383–3390. [CrossRef]

118. Lee, S.; Jeong, Y.; Jeong, S.; Lee, J.; Jeon, M.; Moon, J. Solution-processed ZnO nanoparticle-based semiconductor oxide thin-film transistors. *Superlattice. Microst.* **2008**, *44*, 761–769. [CrossRef]

119. Jun, J.H.; Seong, H.; Cho, K.; Moon, B.M.; Kim, S. Ultraviolet photodetectors based on ZnO nanoparticles. *Ceram. Int.* **2009**, *35*, 2797–2801. [CrossRef]

120. Neshataeva, E.; Kummell, T.; Bacher, G.; Ebbers, A. All-inorganic light emitting device based on ZnO nanoparticles. *Appl. Phys. Lett.* **2009**, *94*, 091115. [CrossRef]

121. Wang, M.Q.; Lian, Y.Q.; Wang, X.G. PPV/PVA/ZnO nanocomposite prepared by complex precursor method and its photovoltaic application. *Curr. Appl. Phys.* **2009**, *9*, 189–194. [CrossRef]

122. Saidaminov, M.I.; Adinolfi, V.; Comin, R.; Abdelhady, A.L.; Peng, W.; Dursun, I.; Yuan, M.; Hoogland, S.; Sargent, E.H.; Bakr, O.M. Planar-integrated single-crystalline perovskite photodetectors. *Nat. Commun.* **2015**, *6*, 8724. [CrossRef]

123. Fu, X.-W.; Liao, Z.-M.; Zhou, Y.-B.; Wu, H.C.; Bie, Y.Q.; Xu, J.; Yu, D.P. Graphene/ZnO nanowire/graphene vertical structure based fast-response ultraviolet photodetector. *Appl. Phys. Lett.* **2012**, *100*, 223114. [CrossRef]

124. Lopez-Sanchez, O.; Lembke, D.; Kayci, M.; Radenovic, A.; Kis, A. Ultrasensitive photodetectors based on monolayer MoS_2. *Nat. Nanotechnol.* **2013**, *8*, 497–501. [CrossRef]

125. Dou, L.T.; Yang, Y.; You, J.B.; Hong, Z.; Chang, W.H.; Li, G.; Yang, Y. Solution-processed hybrid perovskite photodetectors with high detectivity. *Nat. Commun.* **2014**, *5*, 5404. [CrossRef]

126. Han, Y.G.; Wu, G.; Wang, M.; Chen, H. High efficient UV-A photodetectors based on monodispersed ligand-capped $_{TiO2}$ nanocrystals and polyfluorene hybrids. *Polymer* **2010**, *51*, 3736–3743. [CrossRef]

127. Jin, Y.Z.; Wang, J.P.; Sun, B.Q.; Blakesley, J.C.; Greenham, N.C. Solution-processed ultraviolet photodetectors based on colloidal ZnO nanoparticles. *Nano Lett.* **2008**, *8*, 1649–1653. [CrossRef]

128. Zhai, T.Y.; Fang, X.S.; Liao, M.Y.; Xu, X.; Zeng, H.; Yoshio, B.; Golberg, D. A comprehensive review of one-dimensional metal-oxide nanostructure photodetectors. *Sensors* **2009**, *9*, 6504–6529. [CrossRef]

129. Sang, L.W.; Liao, M.Y.; Sumiya, M. A comprehensive review of semiconductor ultraviolet photodetectors: From thin film to one-dimensional nanostructures. *Sensors* **2013**, *13*, 10482–10518. [CrossRef]

130. Hu, L.F.; Yan, J.; Liao, M.Y.; Wu, L.; Fang, X. Ultrahigh external quantum efficiency from thin SnO_2 nanowire ultraviolet photodetectors. *Small* **2011**, *7*, 1012–1017. [CrossRef] [PubMed]

131. Li, L.; Lee, P.S.; Yan, C.Y.; Zhai, T.; Fang, X.; Liao, M.; Koide, Y.; Bando, Y.; Golberg, D. Ultrahigh-performance solar-blind photodetectors based on individual single-crystalline In2Ge2O7 nanobelts. *Adv. Mater.* **2010**, *22*, 5145–5149. [CrossRef] [PubMed]

132. Fang, X.S.; Hu, L.F.; Huo, K.F.; Gao, B.; Zhao, L.; Liao, M.; Chu, P.K.; Bando, Y.; Golberg, D. New ultraviolet photodetector based on individual Nb_2O_5 nanobelts. *Adv. Funct. Mater.* **2011**, *21*, 3907–3915. [CrossRef]

133. Wang, J.X.; Yan, C.Y.; Kang, W.B.; Lee, P.S. High-efficiency transfer of percolating nanowire films for stretchable and transparent photodetectors. *Nanoscale* **2014**, *6*, 10734–10739. [CrossRef]

134. Li, L.D.; Gu, L.L.; Lou, Z.; Shen, G. ZnO quantum dot decorated Zn_2SnO_4 nanowire heterojunction photodetectors with drastic performance enhancement and flexible ultraviolet image sensors. *ACS Nano* **2017**, *11*, 4067–4076. [CrossRef]

135. Wu, L.; Wu, Y.S.; Pan, X.R.; Kong, F. Synthesis of ZnO nanorod and the annealing effect on its photoluminescence property. *Opt. Mater.* **2006**, *28*, 418–422. [CrossRef]
136. Wu, Y.L.; Tok, A.I.Y.; Boey, F.Y.C.; Zeng, X.T.; Zhang, X.H. Surface modification of ZnO nanocrystals. *Appl. Surf. Sci.* **2007**, *253*, 5473–5479. [CrossRef]
137. Qin, L.Q.; Shing, C.; Sawyer, S. Metal-semiconductor-metal ultraviolet photodetectors based on zinc-oxide colloidal nanoparticles. *IEEE Electron Device Lett.* **2011**, *32*, 51–53. [CrossRef]
138. Khokhra, R.; Bharti, B.; Lee, H.-N.; Kumar, R. Visible and UV photo-detection in ZnO nanostructured thin films via simple tuning of solution method. *Sci. Rep.* **2017**, *7*, 15032. [CrossRef]
139. Gogurla, N.; Sinha, A.K.; Santra, S.; Manna, S.; Ray, S.K. Multifunctional Au-ZnO plasmonic nanostructures for enhanced UV photodetector and room temperature NO sensing devices. *Sci. Rep.* **2014**, *4*, 6483. [CrossRef] [PubMed]
140. Ismail, R.A.; Ali, A.K.; Hassoon, K.I. Preparation of a silicon heterojunction photodetector from colloidal indium oxide nanoparticles. *Opt. Laser Technol.* **2013**, *51*, 1–4. [CrossRef]
141. Seong, H.; Cho, K.; Kim, S. Photocurrent characteristics of solutionprocessed HgTe nanoparticle thin films under the illumination of 1.3 μm wavelength light. *Semicond. Sci. Technol.* **2008**, *23*, 075011. [CrossRef]
142. Jana, M.K.; Chithaiah, P.; Murali, B.; Krupanidhi, S.B.; Biswas, K.; Rao, C.N.R. Near infrared detectors based on HgSe and HgCdSe quantum dots generated at the liquid–liquid interface. *J. Mater. Chem. C* **2013**, *1*, 6184–6187. [CrossRef]
143. Chen, M.; Yu, H.; Kershaw, S.V.; Xu, H.; Gupta, S.; Hetsch, F.; Rogach, A.L.; Zhao, N. Fast, air-stable infrared photodetectors based on spraydeposited aqueous HgTe quantum dots. *Adv. Funct. Mater.* **2014**, *24*, 53–59. [CrossRef]
144. Im, S.H.; Kim, H.; Kim, S.W.; Kim, S.-W.; Il Seok, S. Efficient HgTe colloidal quantum dot-sensitized near-infrared photovoltaic cells. *Nanoscale* **2012**, *4*, 1581–1584. [CrossRef]
145. Nam, M.; Kim, S.; Kim, S.; Jeong, S.; Kim, S.-W.; Lee, K. Near-infrared-sensitive bulk heterojunction solar cells using nanostructured hybrid composites of HgTe quantum dots and a low-band gap polymer. *Sol. Energy Mater. Sol. Cells* **2014**, *126*, 163–169. [CrossRef]
146. Lhuillier, E.; Guyot-Sionnest, P. Recent progresses in mid infrared nanocrystal optoelectronics. *IEEE J. Sel. Top. Quantum Electron.* **2017**, *23*, 1–8. [CrossRef]
147. Jia, B.W.; Tan, K.H.; Loke, W.K.; Wicaksono, S.; Lee, K.H.; Yoon, S.F. Monolithic integration of insb photodetector on silicon for mid-infrared silicon photonics. *ACS Photonics* **2018**, *5*, 1512–1520. [CrossRef]
148. Iacovo, A.D.; Venettacci, C.; Colace, L.; Scopa, L.; Foglia, S. PbS colloidal quantum dot visible-blind photodetector for early indoor fire detection. *IEEE Sens. J.* **2017**, *17*, 4454–4459. [CrossRef]
149. Geyer, S.M.; Scherer, J.M.; Jaworski, F.B.; Bawendi, M.G. Multispectral imaging via luminescent down-shifting with colloidal quantum dots. *Opt. Mater. Express* **2013**, *3*, 1167–1175. [CrossRef]
150. Lhuillier, E.; Keuleyan, S.; Zolotavin, P.; Guyot-Sionnest, P. Mid-infrared HgTe/As$_2$S$_3$ field effect transistors and photodetectors. *Adv. Mater.* **2013**, *25*, 137–141. [CrossRef] [PubMed]
151. Chen, M.; Shao, L.; Kershaw, S.V.; Yu, H.; Wang, J.; Rogach, A.L.; Zhao, N. Photocurrent enhancement of HgTe quantum dot photodiodes by plasmonic gold nanorod structures. *ACS Nano* **2014**, *8*, 8208–8216. [CrossRef] [PubMed]
152. Tang, X.; Wu, G.F.; Lai, K.W.C. Plasmon resonance enhanced colloidal HgSe quantum dot filterless narrowband photodetectors for midwave infrared. *J. Mater. Chem. C* **2017**, *5*, 362–369. [CrossRef]
153. Yifat, Y.; Ackerman, M.; Guyot-Sionnest, P. Mid-IR colloidal quantum dot detectors enhanced by optical nano-antennas. *Appl. Phys. Lett.* **2017**, *110*, 041106. [CrossRef]

Review

Low-Cost Microbolometer Type Infrared Detectors

Le Yu, Yaozu Guo, Haoyu Zhu, Mingcheng Luo, Ping Han and Xiaoli Ji *

School of Electronic Science and Engineering, Nanjing University, Nanjing 210093, China; yule@nju.edu.cn (L.Y.); dg1923016@smail.nju.edu.cn (Y.G.); mf1923107@smail.nju.edu.cn (H.Z.); mingchengluo@163.com (M.L.); hanping@nju.edu.cn (P.H.)
* Correspondence: xji@nju.edu.cn; Tel.: +86-025-89683965

Received: 31 July 2020; Accepted: 20 August 2020; Published: 24 August 2020

Abstract: The complementary metal oxide semiconductor (CMOS) microbolometer technology provides a low-cost approach for the long-wave infrared (LWIR) imaging applications. The fabrication of the CMOS-compatible microbolometer infrared focal plane arrays (IRFPAs) is based on the combination of the standard CMOS process and simple post-CMOS micro-electro-mechanical system (MEMS) process. With the technological development, the performance of the commercialized CMOS-compatible microbolometers shows only a small gap with that of the mainstream ones. This paper reviews the basics and recent advances of the CMOS-compatible microbolometer IRFPAs in the aspects of the pixel structure, the read-out integrated circuit (ROIC), the focal plane array, and the vacuum packaging.

Keywords: microbolometer; complementary metal oxide semiconductor (CMOS)-compatible; uncooled infrared detectors; thermal detectors; infrared focal plane array (IRFPA); read-out integrated circuit (ROIC)

1. Introduction

Infrared (IR) detectors are devices that measure the incident IR radiation by turning it into other easy-to-measure physical phenomenon. The IR detectors may be classified into photon detectors and thermal detectors according to the basis of their operating principle [1]. The photon IR detector absorbs the radiation by the interaction with electrons in the semiconductor material, and then the variation in the electronic energy distribution results in observable electrical output signal. This kind of detectors shows perfect signal-to-noise performance and very fast response, while its utilization is limited because of the requirement of cryogenic cooling [2–5]. Compared to its competitor, the thermal IR detector, which absorbs the incident IR power to cause temperature rise and measures the consequent change in some physical properties, presents smaller volume, lower cost, and non-necessity of cryogenic cooling, therefore it has wide application in automobile, security, and electric appliance [6–8]. The development of thermal IR detectors could be traced back to Langley's bolometer in 1880, which use two platinum foils to form the arms of a Wheatstone bridge [9]. However, thermal IR detectors failed to attract sufficient attention until the last decade of the 20th century. The reason is that the thermal IR detectors are considered to be much slower and less insensitive than the photon IR detectors [6]. In 1992, both Texas Instruments and Honeywell published their uncooled IRFPAs (infrared focal plane array) based on pyroelectric type and microbolometer type thermal detector, respectively, with fascinating performance [10,11], successfully encouraging a sustained effort to further reduce the pixel size, improve the device performance, and reduce the production cost [12–46].

Today, one of the most attractive thermal IR detectors for imaging purpose is the microbolometer IRFPA. Comparing to other thermal IR detectors like thermopile detector [47–50], pyroelectric detector [51–54], and superconducting transition edge sensor (TES) bolometer detector [55–58], it is promising for the commercial imaging applications because of its respectable performance, small pixel size, and ease to make [59]. Attributing to the continuous efforts and the technological advances, the pixel size of the microbolometer detector fabricated via the low-cost manufacture technology based on silicon LSI (large scale integration) circuit process has been reduced to beyond 17 μm [18–20]. Not only does the high-integration process lower the production cost of the detectors, but also it provides mature approach with small feature size and high uniformity to benefit the pixel size and the device performance. Especially, the complementary metal oxide semiconductor (CMOS) microbolometer technology is developed for long-wavelength IR (LWIR, 8–14 μm) FPAs via CMOS foundry compatible approaches [23–46]. During the fabrication process, the layer structures of the absorber and the thermal sensor are formed with CMOS process, and then post-CMOS micro-electro-mechanical system (MEMS) process are used to form suspended microbridge structures in purpose of thermal isolation. This technology aims to eliminate the requirement of special process and simplify the post-CMOS MEMS process in order to achieve the ultra-low-cost microbolometer IRFPAs.

However, the most common thermistor materials like vanadium oxide (VO_x) [60–62] and silicon derivatives (a-Si, a-SiGe, a-$Ge_xSi_{1-x}O_y$, etc.,) [63–65], which have appropriate electrical properties, are not compatible with the CMOS process. For the CMOS-compatible microbolometer IR detector, one choice is the p-n junction diode which has acceptable properties and compatibility with CMOS process; therefore the silicon-on-insulator (SOI) diode IRFPAs have attracted continuous attention since first reported by Ishikawa et al. in 1999 [13], and have been widely adopted in low-cost commercial IR detectors. Besides, CMOS-compatible metal or semiconductor materials (e.g., aluminum [41–43], titanium [12,29], polycrystalline silicon [44], etc.,) have been investigated as another choice as well. Although the SOI diode IRFPA and the CMOS-compatible material microbolometer IRFPA have relative low temperature coefficient, it could be compensated by the high integration and high uniformity. Till now, a lot of efforts have been done to improve these two types of microbolometer detectors: Ueno et al. proposed a multi-level structure that has an independent metal reflector between the absorber and the thermistor for interference IR absorption in SOI diode IRFPA [15]; Takamuro et al. invented the 2-in-1 SOI diode pixel technology to significantly increase the diode series number in a pixel, leading to the increase of responsivity [18]; Ning et al. implemented a double-sacrificial-layer aluminum microbolometer fabrication process to enhance both the thermal isolation of the suspended microbridge structure and the IR absorption of the optical resonant cavity [42].

In this paper, we focus on the CMOS-compatible microbolometer IR detectors, that is, the low-cost microbolometer type IR detectors for imaging purpose fabricated via CMOS process (or conventional silicon LSI circuit process). During the fabrication process, no special delicate approach (e.g., the deposition of vanadium oxides) should be needed, and only simple MEMS process is applied after the CMOS process. The basics and the fabrication processes of such low-cost microbolometer IR detectors will be introduced, while the development trends and the technological advances are also discussed.

2. Theory and Development Trends

2.1. Basics of Microbolometer

When the IR radiation falls on the surface of the bolometer, it is absorbed and results in a temperature increase ΔT. When the heat balance is reached, the temperature rise is

$$\Delta T = \frac{\varepsilon P_0}{\left(G^2 + \omega^2 C^2\right)^{1/2}} = \frac{\varepsilon P_0}{G\left(1 + \omega^2 \tau^2\right)^{1/2}} \tag{1}$$

here C is the thermal capacitance of the absorber, which is connected to the environment via the thermal conductance G. ε is the emissivity (absorptance) of the incident IR radiation with amplitude P_0 and angular frequency ω. τ is the thermal time constant, which commonly ranges from several to several tens of milliseconds for the thermal IR detector. For both resistance type and diode type microbolometers, the temperature increase is transferred into the electric signal and then measured.

Lower thermal conductance results in larger temperature increase and higher sensitivity, but worse time constant. Therefore, a small thermal capacitance is always necessary in order to relax the restriction of the trade-off between the sensitivity and the thermal constant time.

The output signal of the microbolometer accompanies with noise that originates from various uncorrelated source, resulting in undesired random fluctuations. There are several major noise sources that should be considered in a microbolometer IR detector: Johnson noise, temperature fluctuation noise, and $1/f$ noise [66]. Besides, the shot noise also could be taken into consideration for the diode type microbolometer detector [27]. The total noise could be calculated in terms of its mean square as the sum of the mean squares of these noises:

$$\overline{V_n^2} = \overline{V_J^2} + \overline{V_{TF}^2} + \overline{V_{1/f}^2} + \overline{V_{shot}^2} \tag{2}$$

These noises determine the noise equivalent temperature difference (NETD). NETD is defined as the change in temperature when the output signal equals to the noise, i.e., the minimum temperature difference that could be measured. The performance of a microbolometer IR detector with optics may be evaluated in terms of the NETD. It is given by [67].

$$NETD = \frac{\left(4F^2 + 1\right)\sqrt{\overline{V_n^2}}}{A\varepsilon R_v (dP/dT_t)_{\lambda_1 - \lambda_2}} \tag{3}$$

Here $F = f/D$ is the F-number of the optical system, where f and D are the focal length and the aperture of the optics, respectively. A is the size of the absorber, R_v is the responsivity defined as the change of the output voltage resulted from per unit incident IR power, $(dP/dT_t)_{\lambda 1 - \lambda 2}$ is the change in power per unit area radiated by a blackbody at temperature T_t measured within the IR spectral band from λ_1 to λ_2. The value of $(dP/dT_t)_{\lambda 1 - \lambda 2}$ for a 295 K blackbody within the 8–14 μm band is 2.62×10^{-4} W/cm^{2}K [68]. The NETD of a low-cost microbolometer IRFPA under its operation condition typically ranges from 50 to 500 mK.

2.2. Development Trends

Before the thermal detector has been demonstrated to be practical for imaging purpose, the IR detector field was dominated by the photon detectors which were restricted to military applications because of the expensive materials and requirement of cryogenic coolers. The appearance of the commercialized thermal IR detector encourages the expectation for non-military application. The CMOS-compatible microbolometer IRFPA, from its very beginning, aims to further lower the cost and the chip size, while maintain an acceptable performance.

Pixel size, as the indicator of the integration level, is the key factor limiting the chip size. The pixel size reduction of the CMOS-compatible IR detectors is shown in Figure 1. Because of the considerable efforts, the pixel size of the SOI diode uncooled IRFPAs has been reduced to 15 μm in 2011 [18]. Meanwhile, the spatial resolution or the array size is related to the pixel size. With the progress of the CMOS microbolometer technology, the array size also increases from 128 × 128 reported in 1996 [12] to 2000 × 1000 reported in 2012 [19].

As shown in Figure 2, unlike the pixel size and spatial resolution, the NETD generally shows a trend of remaining in the same level rather than continuously improving. Since the CMOS microbolometer IR detectors mainly aim the non-military market, a NETD of several tens mK is already capable in handling those applications.

Smaller pixels collect less IR power to increase the temperature, resulting in lower sensitivity. In a conventional structure of the microbolometer pixel, the absorber, the thermal sensor, and the supporting legs are in the same suspended layer. When the pixels are scaled down, higher fill factor and higher emissivity are necessary in maintaining the same sensitivity since less IR radiation is absorbed, bringing a hard task for the trade-off between the thermal conductance and the fill factor. To fix this issue, the multi-level structures with hidden support leg [69–71] or umbrella absorber [15,72,73] have been proposed. By these new structures, the high fill factor and low thermal conductance could be simultaneously achieved in small pixels. However, it seems that the further step of the pixel size reduction has slowed down in recent year, indicating the requirement of novel technical innovation. In addition, the restriction of the diffraction limitation also impedes the progress of pixel size reduction, which is discussed later.

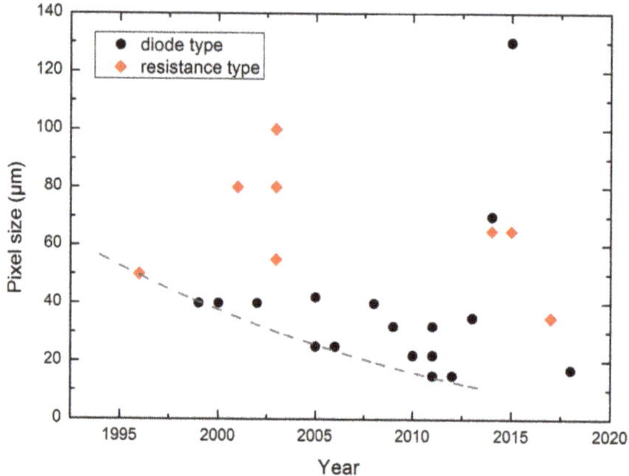

Figure 1. Trends of the pixel size reduction for complementary metal oxide semiconductor (CMOS)-compatible microbolometer infrared (IR) detectors, data taken from (in left-to-right order) [12–15,17–22,25,27–29,31–35,37,39–41,43,46].

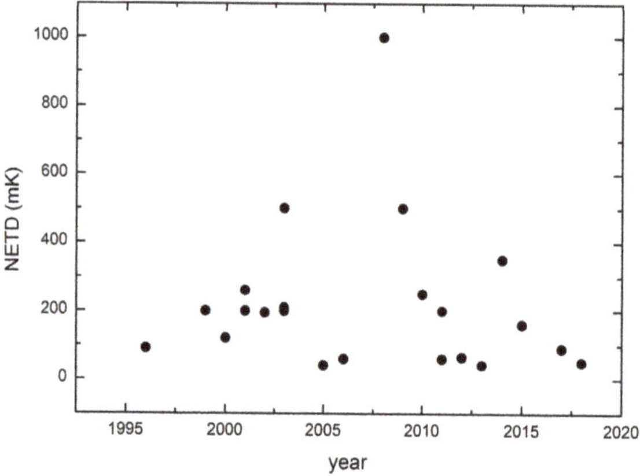

Figure 2. Trends of the noise equivalent temperature difference (NETD) for CMOS-compatible microbolometer IR detectors, data taken from (in left-to-right order) [12–15,17–22,25–29,31,33–35,37,40,46].

3. Complementary Metal Oxide Semiconductor (CMOS)-Compatible Microbolometer Pixel

3.1. The Resistance Type Microbolometer Pixel

Figure 3 shows the pixel structure of the resistance type microbolometer. The microbolometer pixel contains three parts: the infrared absorber, the thermal sensor, and the microbridge structure. The infrared absorber usually consists of the dielectric layer or the multi-layer structure of dielectric and metal layers [74]. The thermal sensor is implemented using a CMOS-compatible thermistor layer sandwiched in the absorber, which is designed to be serpentine to maximize the resistance. The microbridge structure consists of two support legs to sustain the suspended area, creating a thermally isolated cavity between the absorber and the substrate in order to greatly reduce the thermal conductance. In an Al microbolometer, the IR absorber is implemented using the SiO_2/Si_3N_4 layer, with the Al thermistor from the metal interconnect layer Metal 3 sandwiched inside the SiO_2 layer. The SiO_2 and Si_3N_4 also provide protection for the thermistor and the read-out circuit during the post-CMOS etching process.

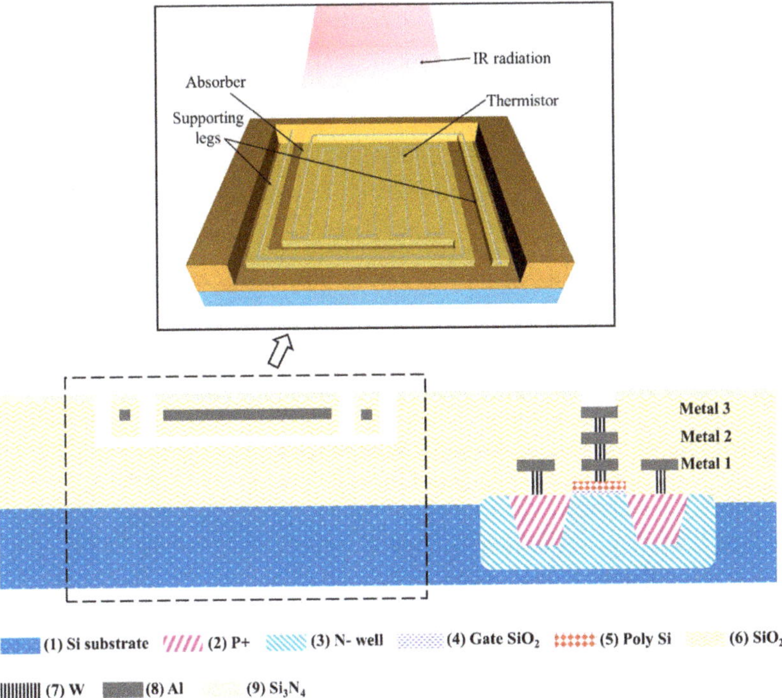

Figure 3. The cross-sectional schematics showing the pixel structure of an aluminum microbolometer. The inset provides a 3D perspective view on the structure of the absorber and the thermistor.

As shown in Figure 4, the process flow of the Al microbolometer shown in Figure 3 is as follows:

a.   The p+/n−well (2,3), gate oxide (4), and polysilicon (5) are fabricated on the substrate (1) via lithography, deposition, ion implantation, and annealing in order to form the transistor.

b.   Deposit SiO_2 (6) as the isolation layer, then etch and deposit W (7) to form the contacts. Afterwards the metal interconnect layer Metal 1 (and the subsequent metal interconnection layers in the active region as well) is formed by depositing Al (8) as the connection of the read-out circuit.

c.   Deposit SiO_2 (6) and then form the W (7) vias. The Al (8) in metal interconnect layer Metal 2 is deposited as the sacrificial layer in the sensor region.

d.   Deposit SiO_2 (6), form the W (7) vias, and then deposit Al (8) for the interconnect layer Metal 3 to form the thermistor in the sensor region.

e.   Deposit SiO_2/Si_3N_4 (6,9) to protect the device. Then dry etch the SiO_2/Si_3N_4 over the pad area and expose the sacrificial layer.

f.   Use photoresist (10) to protect the pad area during the post-CMOS etching. Use the phosphoric acid solution to etch the sacrificial layer to form the cavity and expose the microbridge structure.

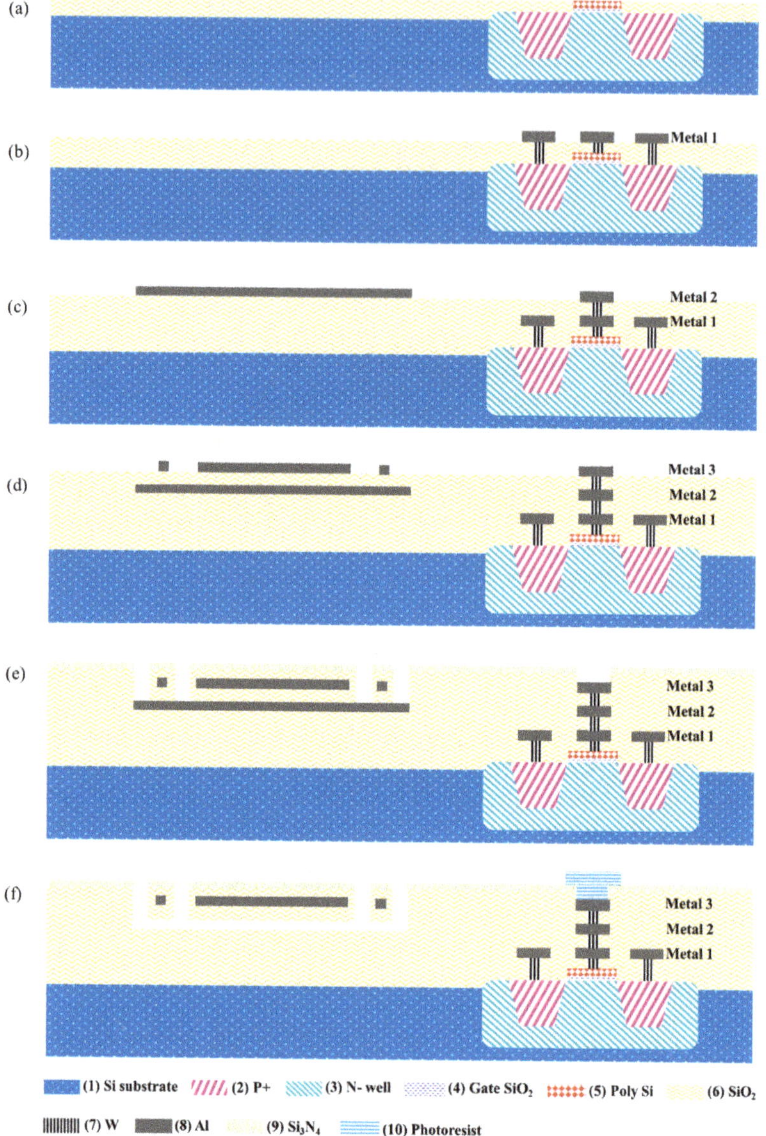

Figure 4. The process flow for an Al microbolometer: (**a–e**) are in a standard CMOS process and (**f**) is in a subtractive micro-electro-mechanical system (MEMS) process.

The steps a to e are in a standard CMOS process, while step f is in a post-CMOS MEMS process. The whole process could be completed in a CMOS foundry to achieve high uniformity devices in ultra-low production cost. However, an intrinsic limitation of the CMOS-compatible microbolometer is the thermistor material. When infrared radiation illuminates the surface of the absorber, the thermistor in the absorber is heated and causes a change in its resistance related to its temperature coefficient of resistance (TCR) α, defined as:

$$\alpha = \frac{1}{R_0} \cdot \frac{dR_b}{dT} \tag{4}$$

R_0 is the resistance of the bolometer at room temperature; dR_b is the resistance change depending on the temperature change dT. Under a certain bias current, the change of the thermistor resistance could be obtained by measuring the output voltage. Therefore, the value of TCR significantly influences the device sensitivity. Generally, the semiconductor-based microbolometers have negative TCR values, while the metal ones have positive TCR values. Table 1 lists several common CMOS-compatible thermistor materials. Compared to the high TCR thermistor materials like VO_x which has a TCR of about 2–3%/K, the CMOS-compatible materials have obvious disadvantage in the TCR. This results in a low sensitivity which needs to be compensated by the high-spec read-out circuit.

Table 1. Resistivity and temperature coefficient of resistance (TCR) of several complementary metal oxide semiconductor (CMOS)-compatible materials [12,75].

Material	Resistivity (10^{-4} $\Omega \cdot$cm, at 300 K)	TCR (%/K)
undoped polysilicon	199	−0.085
n−polysilicon	62.2	−0.016
Al	0.03	0.38
Ti	1.2	0.25

3.2. The Diode Type Microbolometer Pixel

The pixel structure of the diode type microbolometer is similar to that of the resistance type microbolometer; it also consists of three parts: the infrared absorber, the thermal sensor, and the microbridge structure. Here the thermal sensor becomes the p-n junction diodes, which are connected in series to enlarge the output signal. The diodes are usually fabricated on the SOI film for several reasons: (a) The diodes fabricated on deposited Si film exhibit large $1/f$ noise [76,77]; (b) the diodes fabricated on Si substrate need a special electrochemical etch-stop technique to protect the n−well during the post-CMOS etching process [27,33,78]; (c) the SOI film is expected to have fewer defects and localized states which could reduce the $1/f$ noise. The pixel structure of a SOI diode detector is shown in Figure 5. The BOX (buried oxide) layer and the dielectric film over the diodes protect the diodes during the post-CMOS etching process.

The temperature change in the diode under a certain bias current results in a voltage shift. The temperature coefficient in a diode type microbolometer is determined by the forward voltage V_f. With diodes in series connection, the sensitivity is given by [14]:

$$\frac{dV_f}{dT} = -\frac{n(1.21 - V_f/n)}{T} \tag{5}$$

where n is the number of the diodes in the series. The typical value of the sensitivity for a single diode at 300 K is ~2 mV/K under a bias voltage of 0.6 V [59], which is equivalent to a temperature coefficient of only ~0.33%/K. However, as the number of the diodes in the series increases, the temperature coefficient could become comparable to the TCR of VO_x. For instance, when $n = 8$, the diodes in series connection have a temperature coefficient of ~3%/K. Meanwhile, benefiting from the high uniformity of the CMOS process and the low defect density in the SOI film, the diode type microbolometer usually exhibits much better noise.

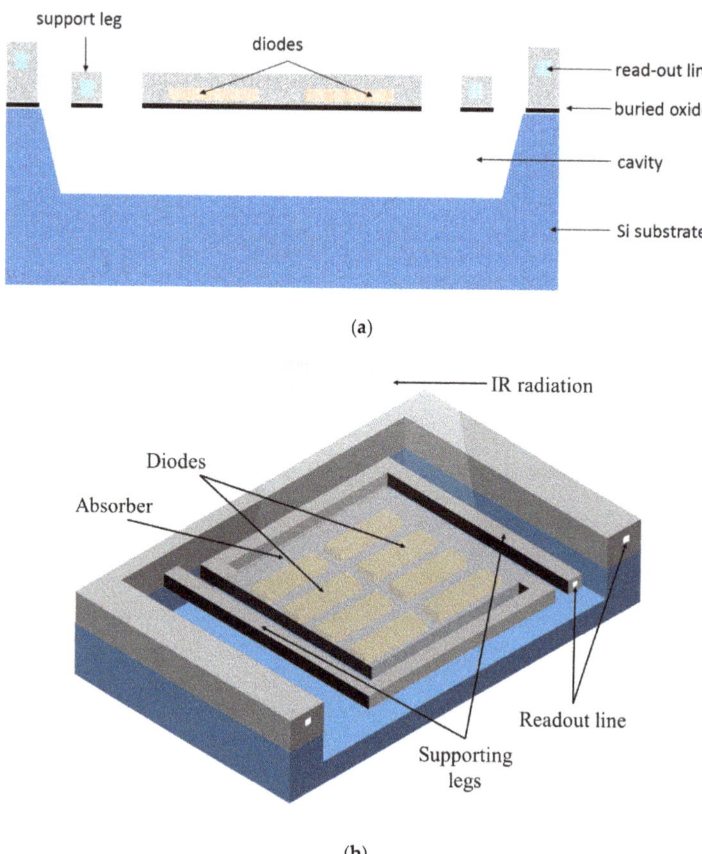

Figure 5. The pixel structure of a diode type microbolometer: (**a**) cross-sectional view and (**b**) 3D perspective view.

3.3. Improvement in Absorber for Small Pixel Structure

The small pixels benefit the detectors from a production point of view. For instance, when the scaling down from 25 μm pixel to 17 μm, it decrease the processing cost by 40% and the power consumption by 33%, while the detection range is increased significantly [61]. However, since the IR absorption is proportional to the absorber area, it demands novel structures that achieve high fill factor or high emissivity in order to compensate the disadvantage of small pixel size.

The umbrella absorber is a widely adopted design to maximize the absorber area that captures more incident IR energy. As shown in Figure 6a, it consists of an IR absorber layer, which individually suspends over the bolometer and support legs, supported by one or more posts. The umbrella absorber consists of dielectric layer or multi-layer structure of metal and dielectric layers, which is the same as that of the conventional absorber layers. Some umbrella absorbers have etch holes designed to enhance the sacrificial removal. These etch holes also benefit the responsivity due to the decrease of thermal capacitance of the umbrella absorber [79]. The umbrella absorber can achieve a fill factor above 90% and ~23% improvement in responsivity [72]. This structure provides the fill factor close to an ideal value at the expense of more process steps, usually increasing 2–5 masking layers and corresponding deposition and etching steps [45].

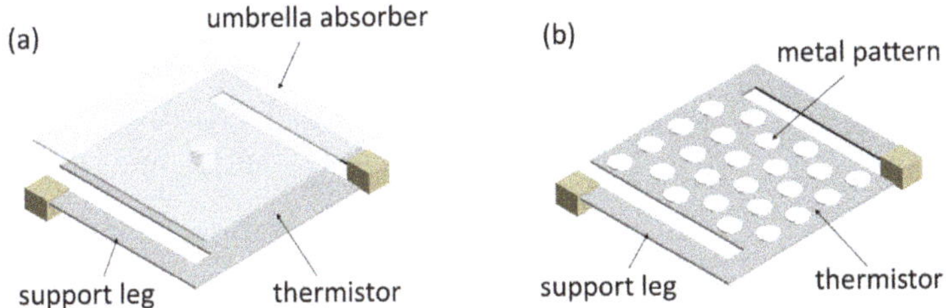

Figure 6. The schematics of a microbolometer with (**a**) umbrella absorber; (**b**) metasurface.

Another prospective approach to improve the absorption is the absorber with a metasurface. The magnetic resonance in the metasurface could control the thermal emission of phonon, therefore the IR absorption spectrum of the metasurface could be manipulated via changing the structure parameter [80]. This could be implemented to the surface of the absorber in order to enhance the IR absorption in the microbolometer pixel, as shown in Figure 6b. This novel approach has attracted the attention of several groups and the preliminary results reveal its potential of frequency selection and absorption enhancement [81–83].

4. Read-Out Integrated Circuit (ROIC)

The IR energy absorbed by the microbolometer pixel is transformed into weak photocurrent, which is not capable for direct processing due to the noise interference. The photocurrent needs to be amplified and finally turned into digital signal by the read-out integrated circuit (ROIC). Benefiting from the CMOS technology, the ROIC has the advantages of high signal handling capacity, high circuit density, low power dissipation, high uniformity and low noise [3]. As shown in Figure 7, the ROIC usually contains several blocks: (1) The read-out circuit (ROC) to amplify the photocurrent and turn it into a voltage signal; (2) the row decoder and the column multiplexer to select an individual pixel; (3) the power supply and clock signal generator to provide the bias and the clock signal; (4) some IRFPAs have the on-chip analog-to-digital converter (ADC) integrated in the ROIC, while others implement the external ADC. Among all these blocks, the ROC and the ADC are the core blocks which determine the performance of the ROIC.

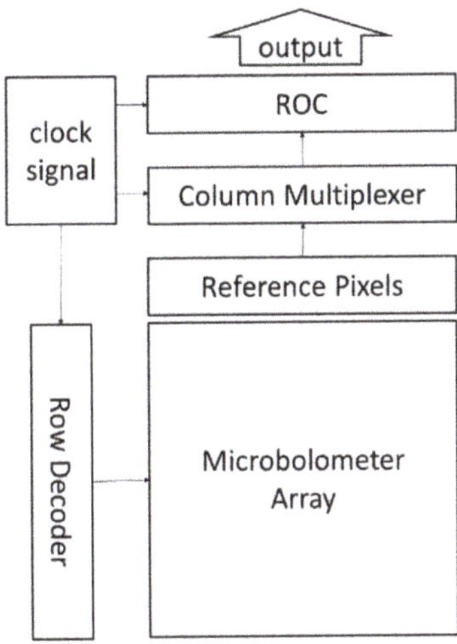

Figure 7. The block schematics of a read-out integrated circuit (ROIC).

4.1. Read-Out Circuit (ROC)

In the ROC, the photocurrent generated from the pixel is amplified and accumulated by a capacitor during an integration time to form a stronger voltage signal, which is then read out into a sample-and-hold (S/H) circuit for the consequent digital conversion in ADC. The design of the ROC significantly affects the power dissipation and the quality of the analog output signal before converting. The most commonly used ROC configuration in microbolometer IRFPAs are direct injection (DI) [61,84,85], gate modulation input (GMI) [13,34,35], and capacitive transimpedance amplifier (CTIA) [20,31,65,86,87]. The design concepts involve the performance and the structural complexity; each designer may prefer a different design depending on the technical requirement and the process schedule.

The structure of the DI configuration is shown in Figure 8a. The photocurrent is injected to C1 to integrate after being amplified via M1, and then is read out to S/H circuit through M4. The function of M2 is to reset the voltage on C1. The DI benefits from simple structure and low power dissipation, but suffers from unstable bias voltage, poor linearity, and poor noise suppression. Figure 8b shows the structure of the GMI configuration. The photocurrent flows into a current-mirror to generate the mirror current toward C1 and then gets integrated. The GMI itself has a varying current gain depending on the background, therefore leading to the higher sensitivity, background suppression, and high dynamic range. Meanwhile, the circuit noise is suppressed by the current mirror structure. The disadvantage of GMI is that the linearity is still affected by the unstable bias voltage, while the current gain and injection efficiency are susceptible to the threshold voltage and process condition of the metal-oxide-semiconductor field-effect transistor (MOSFET), resulting to a negative influence on the circuit performance.

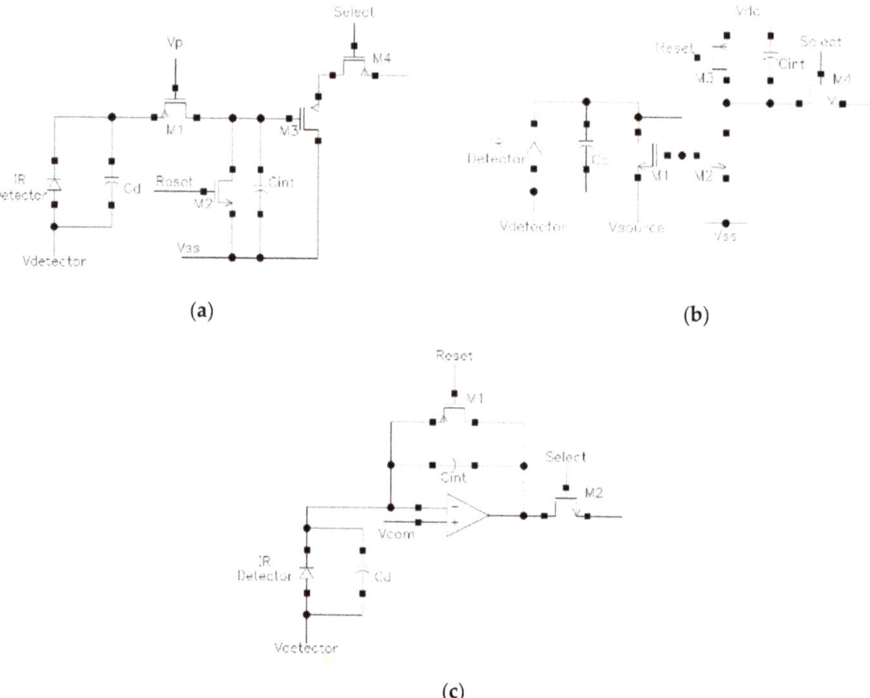

(a)

(b)

(c)

Figure 8. Structures of read-out circuit configuration: (**a**) direct injection (DI), (**b**) gate modulation input (GMI), and (**c**) capacitive transimpedance amplifier (CTIA).

As the most popular configuration in microbolometer IRFPAs, the CTIA configuration is shown in Figure 8c, which is an integrator with the capacitor C1 is in the negative feedback loop of the operational amplifier. M1 is the reset switch and M2 controls the output. The CTIA has low input impedance thus high injection efficiency, stable bias thus excellent linearity, controllable current gain, high sensitivity, and good jam-proof. However, it has relatively high power dissipation, large occupied area, and would introduce more noise due to the offset voltage. Compared to the DI configuration, the CTIA has higher current gain which provides higher sensitivity to detect weaker current, and it also has lower input impedance leading to higher injection efficiency. Compared to the GMI configuration, the CTIA provides more stable bias voltage for the detector, resulting in a better linearity in the output signal. The typical CTIA parameters for microbolometer IRFPAs are shown in Table 2.

Table 2. Parameters of capacitive transimpedance amplifier (CTIA) read-out circuit in microbolometer infrared focal plane arrays (IRFPAs).

Reference	Analog Output Swing (V)	Power Dissipation (mW)	Integration Capacitance (pF)	Linearity	Supply Voltage (V)
METU [87]	2.5	85	1–32 (programmable)		3.3
ULIS [88]	2.8	150		≈1%	5
ULIS [89]	2.8	150		<1%	5
AUT [90]		351			3.3
WPU [91]	2.7	29.8	10		5

4.2. Analog-to-Digital Convertor (ADC)

Generally, the high-speed ADC with a high dynamic range is required for the utility in the CMOS microbolometer IRFPAs. Although the on-chip ADCs using the pixel-level Sigma-Delta (Σ-Δ) ADC [92–94], the monolithic pipeline ADC [95,96], and the column-parallel successive approximation register (SAR) ADC [97] are reported to be available to achieve the high sensitivity ROICs for microbolometer IRFPAs, there are no report about the on-chip ADC for readily available CMOS microbolometer IRFPAs. The CMOS microbolometer IRFPAs usually use external ADCs, due to the inadequate signal processing area in the monolithic FPA. The microbolometer IRFPAs raises the requirement to the ADC such as low power dissipation, high speed, low delay, low offset voltage, low noise, and high slew rate. Table 3 shows typical parameters of an on-chip monolithic pipeline ADC for the microbolometer IRFPA.

Table 3. Parameters of a 14 bits on-chip pipeline analog-to-digital convertor (ADC) designed for microbolometer IRFPA [95].

Effective Number of Bits	Signal-to-Noise Ratio	Differential Nonlinearity	Integral Nonlinearity	Total Harmonic Distortion	Power Consumption
12.9	79 dB	0.76	4.9	−72 dB	95 mW

5. Focal Plane Array (FPA)

Microbolometer pixels are usually fabricated on the substrate with repeating arrangement to form a microbolometer array for imaging purpose. Each microbolometer absorbs the incident IR radiation and transforms it into electric output, which is read out and calibrated by the ROIC to produce a pixel in a two-dimensional image. A microbolometer FPA is the combination of the microbolometer array and the ROIC. Generally, the IRFPA could be sorted as hybrid and monolithic [98]. In the hybrid FPA, the detector pixels and the ROIC are fabricated in different substrates, which are combined using the flip-chip bonding via metal bumps. Since it has the advantages such as the independent optimization of detector material and multiplexer, near 100% fill factor, and sufficient signal processing area, it is widely used in the cooled IRFPAs and high-end uncooled IRFPAs [6]. The monolithic FPA integrates the ROIC and the detector pixels in the same substrate, and part of the column or row selecting circuit is integrated in the pixels. Since the silicon-based monolithic FPA technology is compatible with CMOS process, providing a mature approach with high uniformity and low cost, it is widely used in the microbolometer IRFPAs.

The reduction of pixel size makes challenging tasks for the mechanical stability of the pixel structure, the ROIC, the signal to noise ratio, etc. Not only the thermal sensor material, but also the overall process becomes the limits of the final performance of the IRFPAs. Table 4 lists the performance of several commercial IRFPAs. The performance of SOI diode IRFPAs and the CMOS-compatible resistance microbolometer IRFPAs is still inferior to that of the VO_x or Si derivatives microbolometer, but the gap between the two is small. This means the low sensitivity resulted from the low TCR of the thermal sensor material could be partly compensated by the small feature size and high uniformity provided through CMOS or Si LSI process.

Table 4. CMOS-compatible microbolometer IRFPAs versus other microbolometer IRFPAs.

Reference	Material	Array Size	Pixel Size	ROIC Type	Frame Rate [1]	NETD
Mitsubishi [20]	Diode	320 × 240	17 μm	CTIA	60 Hz	50 mK
MikroSens [45]	CMOS-compatible resistance	120 × 160	35 μm	CTIA	11 Hz	117 mK
Toshiba [37]	Diode	320 × 240	22 μm	GMI	25 ms	200 mK
Raytheon [60]	VO_x	640 × 512	20 μm		30 Hz	<50 mK
DRS [61]	VO_x	1024 × 768	17 μm	DI	30 Hz	<50 mK
FLIR [99]	VO_x	640 × 512	12 μm		60 Hz	<40 mK
L-3 Communications [85]	a-Si/a-SiGe	1024 × 768	17 μm	DI	10 ms	35 mK
ULIS [65]	a-Si	1024 × 768	17 μm	CTIA	30 Hz	46 mK

[1] Or time constant in case the frame rate is not mentioned.

6. Vacuum Packaging Technology

The thermal conduction via the atmosphere takes over a large fraction in the total thermal conduction, especially when the pixel size is small. Since the temperature change and thus the responsivity is proportional to the thermal conductance, the vacuum packaging of the microbolometer pixels is necessary to eliminate the thermal conduction through air. Unfortunately, the cost of the vacuum packaging is one of the major cost drivers for the microbolometer IRFPA. The typical vacuum level required here is below 1 Pa, which raises a challenge to the packaging technology [100]. Although such requirement could be achieved via one-by-one pumping through a fine-bore tube, the cost becomes a bottleneck in lowering the cost of uncooled IRFPAs. Figure 9 shows the concept of the wafer-level packaging (WLP) technology for IRFPA, which is a popular option for cost reduction [59,101,102]. In this technology an IR transparent cap wafer is bonded to the IRFPA wafer under vacuum and then the hermetical sealing is achieved using solders. Several steps are needed prior to the bonding to accomplish the cap wafer. The cavities for the pixels are formed via etching, and then both sides of the cap wafer are antireflection-coated, afterwards the vacuum getters are deposited inside the cavities. The WLP technology is a practical technology that is capable to reach an average seal yield > 95% with correct parameters [103].

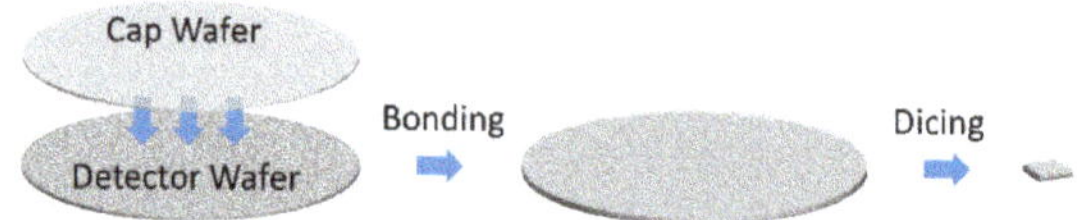

Figure 9. Concept of the wafer level packaging technology.

Although the wafer level packaging technology provides a significant cost reduction, it still takes a considerable proportion in the total cost of the uncooled IRFPA, especially for the low-end market. A pixel level packaging (PLP) technology has been developed to address this issue [104–106]. The PLP process consists in the manufacturing of IR transparent microcaps that cover each pixel in the direct consequent step of the wafer level bolometer fabrication, i.e., no extra bonding process is needed. Figure 10 shows the schematics of a packaged pixel. To form this structure, first, a sacrificial layer with trenches around each pixel is formed above the microbolometer via deposition and etching. Then, an IR transparent material is deposited to form the microcap structure. After that, etch holes are formed through the IR transparent microcap and the sacrificial layer is removed. Finally, a sealing and anti-reflecting layer is deposited under high vacuum. The pixel using PLP keeps a stable vacuum level below 10^{-3} mbar and shows nominal performance after one year of ageing, demonstrating the PLP to be a prospective novel vacuum packaging technology for the microbolometer IRFPAs.

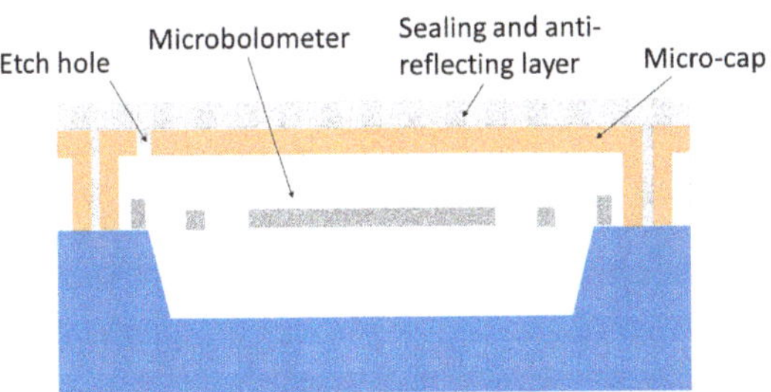

Figure 10. Schematics of the pixel level packaging technology.

7. Limitation and Future Trends

The minimum resolvable size x decided by the diffraction limitation could be expressed by the F-number and the wavelength λ according to the Rayleigh Criterion, which is

$$x \approx f\theta = 1.22\lambda F \tag{6}$$

Here θ is the diffraction angle, and f is the focal length of the optical lens. In a LWIR detector the λ ranges from 8–14 μm, while the F-number for CMOS microbolometer IRFPAs is usually close to 1 to make the device compact, indicating the minimum resolvable size is 10–17 μm. When the pixel size is between $0.5\lambda F$ and $1.22\lambda F$, the resolution still benefits from the oversampling but saturates quickly as the pixel size is smaller [79]. However, unlike the photon detectors which prefer a pixel size close to or even smaller than the diffraction limit to achieve the maximum performance, the reported CMOS microbolometer IRFPAs are still in the "detector limit" regime, i.e., still far from the potential limiting performance.

The main factor that limits the pixel size reduction in CMOS microbolometer is the responsivity. As mentioned above, smaller pixel means less IR absorption, resulting in low responsivity. Meanwhile, the scale-down of circuits results in a lower applied bias voltage, which also means lower responsivity. The responsivity could be enhanced by adjusting the fill factor, the emissivity ε, the thermal conductance G, and the temperature coefficient TCR or dV_f/dT. The fill factor and the emissivity in the state-of-the-art technology are already high, although the ε is still capable to increase to a certain extent via the metasurface technology. The thermal conductance could be decreased with thinner or longer support legs. The TCR is intrinsic to the material, but the resistance increase of the thermistor is able to raise the responsivity. On the other hand, the temperature coefficient of the diode type microbolometer is mainly determined by the number of diodes in series. In any case, the way to enhance the responsivity of the small pixels is related to a smaller feature size.

Besides, the spatial resolution is also affected by the array size. Although the XGA (extended graphics array) format (1024×768) is popularized in the VO_x and silicon derivatives microbolometer IRFPAs, the QVGA (quarter video graphics array) format (320×240) is still popular with the CMOS microbolometer IRFPAs. Since the difficulty to achieve larger array size is much easier compared to that to the pixel size reduction, the status of the low spatial resolution could be considered as a trade-off between the production cost and the performance. It also implies that the market demand to the performance improvement in low-end IR detector is not eager. However, the merit of the pixel size reduction is significant. The small pixel provides low production cost, high spatial resolution, and small device size. Although the steps of the pixel size reduction in the CMOS microbolometer IRFPAs has slowed down in recent years because of insufficient market demand, the smaller pixels with lower costs and better performance will come sooner or later as the technology based on smaller feature size becomes practical.

Author Contributions: Conceptualization, X.J., P.H.; writing—original draft preparation, L.Y.; writing—review and editing, X.J., H.Z., Y.G. M.L.; visualization, L.Y., Y.G. All authors have read and agreed to the published version of the manuscript.

Funding: This research was funded by National Key Research and Development Program of China (2016YFB0400402, 2016YFA0202102, 2016YFB0402403); Key Laboratory of Infrared Imaging Materials and Detectors (IIMDKFJJ19-07).

Conflicts of Interest: The authors declare no conflict of interest. The funders had no role in the design of the study; in the collection, analyses, or interpretation of data; in the writing of the manuscript, or in the decision to publish the results.

References

1. Rogalski, A. *Infrared Detectors*; CRC Press: Boca Raton, FL, USA, 2010.
2. Rogalski, A. HgCdTe infrared detector material: History, status and outlook. *Rep. Prog. Phys.* **2005**, *68*, 2267–2336. [CrossRef]
3. Bhan, R.K.; Dhar, V. Recent infrared detector technologies, applications, trends and development of HgCdTe based cooled infrared focal plane arrays and their characterization. *Opto Electron. Rev.* **2019**, *27*, 174–193. [CrossRef]
4. Martyniuk, P.; Rogalski, A. Quantum-dot infrared photodetectors: Status and outlook. *Prog. Quantum Electron.* **2008**, *32*, 89–120. [CrossRef]
5. Rogalski, A. Quantum well photoconductors in infrared detector technology. *J. Appl. Phys.* **2003**, *93*, 4355–4391. [CrossRef]
6. Rogalski, A. Infrared detectors: An overview. *Infrared Phys. Technol.* **2002**, *43*, 187–210. [CrossRef]
7. Crastes, A.; Tissot, J.-L.; Guimond, Y.; Antonello, P.C.; Leleve, J.; Lenz, H.-J.; Potet, P.; Yon, J.-J. Low cost uncooled IRFPA and molded IR lenses for enhanced driver vision. *Opt. Syst. Des.* **2004**, *5251*. [CrossRef]
8. Pregowski, P. Infrared detector arrays in the detection, automation and robotics-trends and development perspectives. *Meas. Autom. Monit.* **2018**, *64*, 82–89.
9. Barr, E.S. The infrared pioneers—III. *Samuel Pierpont Langley. Infrared Phys.* **1963**, *3*, 195–206. [CrossRef]
10. Hanson, C.M.; Beratan, H.R.; Owen, R.A.; Corbin, M.; McKenney, S. Uncooled thermal imaging at Texas Instruments. In Proceedings of the SPIE's 1993 International Symposium on Optics, Imaging, and Instrumentation, San Diego, CA, USA, 11–16 July 1993.
11. Wood, R.; Han, C.; Kruse, P. Integrated uncooled infrared detector imaging arrays. In Proceedings of the Technical Digest IEEE Solid-State Sensor and Actuator Workshop, Hilton Head Island, SC, USA, 22–25 June 1992; pp. 132–135.
12. Tanaka, A.; Matsumoto, S.; Tsukamoto, N.; Itoh, S.; Chiba, K.; Endoh, T.; Nakazato, A.; Okuyama, K.; Kumazawa, Y.; Hijikawa, M. Infrared focal plane array incorporating silicon IC process compatible bolometer. *IEEE Trans. Electron Devices* **1996**, *43*, 1844–1850. [CrossRef]
13. Ishikawa, T.; Ueno, M.; Endo, K.; Nakaki, Y.; Hata, H.; Sone, T.; Kimata, M. Low-cost 320×240 uncooled IRFPA using conventional silicon IC process. *Opto Electron. Rev.* **1999**, 297–303. [CrossRef]
14. Ishikawa, T.; Ueno, M.; Nakaki, Y.; Endo, K.; Ohta, Y.; Nakanishi, J.; Kosasayama, Y.; Yagi, H.; Sone, T.; Kimata, M. Performance of 320×240 uncooled IRFPA with SOI diode detectors. In Proceedings of the SPIE International Symposium on Optical Science and Technology, San Diego, CA, USA, December 2000; Volume 4130, pp. 152–159.
15. Ueno, M.; Kosasayama, Y.; Sugino, T.; Nakaki, Y.; Fujii, Y.; Inoue, H.; Kama, K.; Seto, T.; Takeda, M.; Kimata, M. 640×480 pixel uncooled infrared FPA with SOI diode detectors. In Proceedings of the SPIE Defense and Security, Orlando, FL, USA, 30 March–1 April 2005; Volume 5783, pp. 566–577.

16. Kimata, M.; Ueno, M.; Takeda, M.; Seto, T. SOI diode uncooled infrared focal plane arrays. In Proceedings of the SPIE Integrated Optoelectronic Devices, San Jose, CA, USA, 1 March 2006.

17. Seto, T.; Kama, K.; Kimata, M.; Takeda, M.; Hata, H.; Nakaki, Y.; Inoue, H.; Kosasayama, Y.; Ohta, Y.; Fukumoto, H. 160 × 120 Uncooled IRFPA for Small JR Camera. In Proceedings of the 5th IEEE Conference on Sensors, Daegu, South Korea, 22–25 October 2006; IEEE: Piscataway, NJ, USA, 2006; pp. 30–33.

18. Takamuro, D.; Maegawa, T.; Sugino, T.; Kosasayama, Y.; Ohnakado, T.; Hata, H.; Ueno, M.; Fukumoto, H.; Ishida, K.; Katayama, H. Development of new SOI diode structure for beyond 17 μm pixel pitch SOI diode uncooled IRFPAs. In Proceedings of the SPIE Defense, Security, and Sensing, Orlando, FL, USA, 25–29 April 2011; Volume 8012.

19. Fujisawa, D.; Maegawa, T.; Ohta, Y.; Kosasayama, Y.; Ohnakado, T.; Hata, H.; Ueno, M.; Ohji, H.; Sato, R.; Katayama, H. Two-million-pixel SOI diode uncooled IRFPA with 15 μm pixel pitch. In Proceedings of the SPIE Defense, Security, and Sensing, Baltimore, MD, USA, 25–26 April 2012; Volume 8353.

20. Fujisawa, D.; Kosasayama, Y.; Takikawa, T.; Hata, H.; Takenaga, T.; Satake, T.; Yamashita, K.; Suzuki, D. Implementation of SOI diode uncooled IRFPA in TEC-less and shutter-less operation. In Proceedings of the SPIE Defense + Security, Orlando, FL, USA, 15–19 April 2018; Volume 10624.

21. Jiang, W.; Ou, W.; Ming, A.; Liu, Z.; Zhang, X. Design and analysis of a high fill-factor SOI diode uncooled infrared focal plane array. *J. Micromech. Microeng.* **2013**, *23*, 065004. [CrossRef]

22. Feng, F.; Zhu, H.; Liu, M.; Wei, X.; Ge, X.; Wang, Y.; Li, X. Uncooled infrared detector based on silicon diode Wheatstone bridge. *Microsyst. Technol. Micro Nanosyst. Inf. Storage Process. Syst.* **2017**, *23*, 669–676. [CrossRef]

23. Tezcan, D.; Kocer, F.; Akin, T. An uncooled microbolometer infrared detector in any standard CMOS technology. In Proceedings of the International Conference On Solid-State Sensors&Actuators (TRANSDUCERS'99), Sendai, Japan, 7–10 June 1999; pp. 7–10.

24. Liu, C.; Mastrangelo, C.H. A CMOS uncooled heat-balancing infrared imager. *IEEE J. Solid State Circ.* **2000**, *35*, 527–535.

25. Tezcan, D.S.; Eminoglu, S.; Akar, O.S.; Akin, T. A low cost uncooled infrared microbolometer focal plane array using the CMOS n-well layer. In Proceedings of the International Conference on Micro Electro Mechanical Systems, Interlaken, Switzerland, 25–25 June 2001; pp. 566–569.

26. Tezcan, D.S.; Eminoglu, S.; Akar, O.S.; Akin, T. Uncooled microbolometer infrared focal plane array in standard CMOS. In Proceedings of the SPIE Symposium on Integrated Optics, San Jose, CA, USA, 15 May 2001; Volume 4288, pp. 112–121.

27. Eminoglu, S.; Tanrikulu, M.Y.; Tezcan, D.S.; Akin, T. Low-cost small-pixel uncooled infrared detector for large focal plane arrays using a standard CMOS process. In Proceedings of the the SPIE AeroSense 2002, Orlando, FL, USA, 1–5 April 2002; Volume 4721, pp. 111–121.

28. Eminoglu, S.; Tezcan, D.S.; Tanrikulu, M.Y.; Akin, T. Low-cost uncooled infrared detectors in CMOS process. *Sens. Actuators A Phys.* **2003**, *109*, 102–113. [CrossRef]

29. Mansi, M.V.; Brookfield, M.; Porter, S.G.; Edwards, I.; Bold, B.; Shannon, J.; Lambkin, P.; Mathewson, A. AUTHENTIC: A very low-cost infrared detector and camera system. In Proceedings of the SPIE International Symposium on Optical Science and Technology, Seattle, WA, USA, 7–11 July 2002; Volume 4820, pp. 227–238.

30. Tezcan, D.S.; Eminoglu, S.; Akin, T. A low-cost uncooled infrared microbolometer detector in standard CMOS technology. *IEEE Trans. Electron Devices* **2003**, *50*, 494–502. [CrossRef]

31. Socher, E.; Sinai, Y.; Nemirovsky, Y. A low-cost CMOS compatible serpentine-structured polysilicon-based microbolometer array. In Proceedings of the TRANSDUCERS'03, 12th International Conference on Solid-State Sensors, Actuators and Microsystems, Digest of Technical Papers (Cat. No. 03TH8664), Boston, MA, USA, 8–12 June 2003; Volume 1, pp. 320–323.

32. CHEN, E.-Z.; Liang, P.-Z. Infrared microbolometer of lateral polysilicon p (+) p (-) n (+) junction based on standard CMOS processes. *J. Infrared Millim. Waves* **2005**, *3*, 227–230.

33. Eminoglu, S.; Tanrikulu, M.Y.; Akin, T. A low-cost 128 × 128 uncooled infrared detector array in CMOS process. *J. Microelectromech. Syst.* **2008**, *17*, 20–30. [CrossRef]

34. Funaki, H.; Honda, H.; Fujiwara, I.; Yagi, H.; Ishii, K.; Sasaki, K. A 160 × 120 pixel uncooled TEC-less infrared radiation focal plane array on a standard ceramic package. In Proceedings of the SPIE Defense, Security, and Sensing, Orlando, FL, USA, 13 April 2009; Volume 7298.

35. Honda, H.; Funaki, H.; Fujiwara, I.; Yagi, H.; Ishii, K.; Suzuki, K.; Sasaki, K.; Ogata, M.; Ueno, R.; Kwon, H. A 320 × 240 pixel uncooled TEC-less infrared radiation focal plane array with the reset noise canceling algorithm. In Proceedings of the SPIE Defense, Security, and Sensing, Orlando, FL, USA, 5–9 April 2010; Volume 7660.

36. Ueno, R.; Honda, H.; Kwon, H.; Ishii, K.; Ogata, M.; Yagi, H.; Fujiwara, I.; Suzuki, K.; Sasaki, K.; Funaki, H. Uncooled Infrared Radiation Focal Plane Array with Low Noise Pixel Driving Circuit. *IEICE Trans. Electron.* **2010**, *93*, 1577–1582. [CrossRef]

37. Fujiwara, I.; Sasaki, K.; Suzuki, K.; Yagi, H.; Kwon, H.; Honda, H.; Ishii, K.; Ogata, M.; Atsuta, M.; Ueno, R. Scale down of pn junction diodes of an uncooled IR-FPA for improvement of the sensitivity and thermal time response by 0.13-μm CMOS technology. In Proceedings of the SPIE Defense, Security, and Sensing, Orlando, FL, USA, 25–29 April 2011; Volume 8012.

38. Kwon, H.; Suzuki, K.; Ishii, K.; Yagi, H.; Honda, H.; Atsuta, M.; Fujiwara, I.; Sasaki, K.; Ogata, M.; Ueno, R. A SOI-based CMOS-MEMS IR image sensor with partially released reference pixels. *J. Micromech. Microeng.* **2011**, *21*, 025028. [CrossRef]

39. Weibing, W.; Dapeng, C.; Anjie, M.; Wen, O.; Zhanfeng, L. Integration of uncooled diode infrared focal plane array. *Infrared Laser Eng.* **2011**, *40*, 997–1000.

40. Tepegoz, M.; Kucukkomurler, A.; Tankut, F.; Eminoglu, S.; Akin, T. A miniature low-cost LWIR camera with a 160 × 120 microbolometer FPA. In Proceedings of the SPIE Defense + Security, Baltimore, MD, USA, 6–8 May 2014; Volume 9070.

41. Shen, N.; Tang, Z.; Yu, J.; Huang, Z. A low-cost infrared absorbing structure for an uncooled infrared detector in a standard CMOS process. *J. Semicond.* **2014**, *35*, 034014. [CrossRef]

42. Shen, N.; Huang, Z.; Tang, Z. A double-sacrificial-layer uncooled infrared microbolometer in a standard CMOS process. In Proceedings of the TENCON 2015-2015 IEEE region 10 conference, Macao, China, 1–4 November 2015; pp. 1–5.

43. Shen, N.; Yu, J.; Tang, Z. An uncooled infrared microbolometer array for low-cost applications. *IEEE Photonics Technol. Lett.* **2015**, *27*, 1247–1249. [CrossRef]

44. Cao, X.; Luo, M.; Si, W.; Yan, F.; Ji, X. Design of microbolometer by polycrystalline silicon in standard CMOS technology. In Proceedings of the China Semiconductor Technology International Conference, Shanghai, China, 18–19 March 2019.

45. Akin, T. Low-cost LWIR-band CMOS infrared (CIR) microbolometers for high volume applications. In Proceedings of the 2020 IEEE 33rd International Conference on Micro Electro Mechanical Systems (MEMS), Vancouver, BC, Canada, 18–22 June 2020; pp. 147–152.

46. Tankut, F.; Cologlu, M.H.; Askar, H.; Ozturk, H.; Dumanli, H.K.; Oruc, F.; Tilkioglu, B.; Ugur, B.; Akar, O.S.; Tepegoz, M. An 80 × 80 microbolometer type thermal imaging sensor using the LWIR-band CMOS infrared (CIR) technology. In Proceedings of the SPIE Defense + Security, Anaheim, CA, USA, 9–13 April 2017; Volume 10177.

47. Foote, M.; Gaalema, S. Progress toward high-performance thermopile imaging arrays. In Proceedings of the SPIE Aerospace/Defense Sensing, Simulation, and Controls, Orlando, FL, USA, 16–20 April 2001; Volume 4369.

48. Xu, D.; Xiong, B.; Wang, Y. Modeling of front-etched micromachined thermopile IR detector by CMOS technology. *J. Microelectromech. Syst.* **2010**, *19*, 1331–1340. [CrossRef]

49. Schaufelbuhl, A.; Schneeberger, N.; Munch, U.; Waelti, M.; Paul, O.; Brand, O.; Baltes, H.; Menolfi, C.; Huang, Q.; Doering, E. Uncooled low-cost thermal imager based on micromachined CMOS integrated sensor array. *IEEE/ASME J. Microelectromech. Syst.* **2001**, *10*, 503–510. [CrossRef]

50. Xu, D.; Xiong, B.; Wang, Y. Micromachined thermopile IR detector module with high performance. *IEEE Photonics Technol. Lett.* **2011**, *23*, 149–151. [CrossRef]

51. Holden, A.J. Applications of pyroelectric materials in array-based detectors. *IEEE Trans. Ultrason. Ferroelectr. Freq. Control* **2011**, *58*, 1981–1987. [CrossRef]

52. Stenger, V.; Shnider, M.; Sriram, S.; Dooley, D.; Stout, M. Thin Film Lithium Tantalate (TFLT) Pyroelectric Detectors. In Proceedings of the SPIE, San Francisco, CA, USA, 22–25 January 2012; Volume 8261.

53. Suen, J.Y.; Fan, K.; Montoya, J.; Bingham, C.; Stenger, V.; Sriram, S.; Padilla, W.J. Multifunctional metamaterial pyroelectric infrared detectors. *Optica* **2017**, *4*, 276–279. [CrossRef]

54. Holden, A.J. Pyroelectric sensor arrays for detection and thermal imaging. In Proceedings of the SPIE Defense, Security, and Sensing, Baltimore, MD, USA, 29 April–3 May2013; Volume 8704.

55. Benford, D.; Allen, C.; Kutyrev, A.; Moseley, S.; Shafer, R.; Chervenak, J.; Grossman, E.; Irwin, K.; Martinis, J.; Reintsema, C. Superconducting transition edge sensor bolometer arrays for submillimeter astronomy. In Proceedings of the 11th Intl Symp. on Space Terahertz Tech., Ann Arbor, MI, USA, 1–3 May 2000.

56. Lee, A.T.; Richards, P.L.; Nam, S.W.; Cabrera, B.; Irwin, K.D. A superconducting bolometer with strong electrothermal feedback. *Appl. Phys. Lett.* **1996**, *69*, 1801–1803. [CrossRef]

57. Dominic, J.B.; Johannes, G.S.; Gordon, J.S.; Lyman, P.; Moseley, J.S.H.; Kent, D.I.; James, A.C.; Christine, A.A. Design and fabrication of two-dimensional superconducting bolometer arrays. In Proceedings of the SPIE Astronomical Telescopes + Instrumentation, Glasgow, UK, 21–25 June 2004; Volume 5498.

58. Zmuidzinas, J.; Richards, P.L. Superconducting detectors and mixers for millimeter and submillimeter astrophysics. *Proc. IEEE* **2004**, *92*, 1597–1616. [CrossRef]

59. Kimata, M. Uncooled infrared focal plane arrays. *IEEJ Trans. Electr. Electron. Eng.* **2018**, *13*, 4–12. [CrossRef]

60. Murphy, D.; Ray, M.; Wyles, J.; Hewitt, C.; Wyles, R.; Gordon, E.; Almada, K.; Sessler, T.; Baur, S.; van Lue, D.; et al. 640 × 512 17 µm microbolometer FPA and sensor development. In Proceedings of the SPIE Defense and Security Symposium, Orlando, FL, USA, 10–12 April 2007; Volume 6542.

61. Li, C.; Han, C.-J.; Skidmore, G. Overview of DRS uncooled VOx infrared detector development. *Opt. Eng.* **2011**, *50*, 061017. [CrossRef]

62. Smith, E.M.; Panjwani, D.; Ginn, J.; Warren, A.P.; Long, C.; Figuieredo, P.; Smith, C.; Nath, J.; Perlstein, J.; Walter, N. Dual band sensitivity enhancements of a VO x microbolometer array using a patterned gold black absorber. *Appl. Opt.* **2016**, *55*, 2071–2078. [CrossRef] [PubMed]

63. Tissot, J.-L.; Rothan, F.; Vedel, C.; Vilain, M.; Yon, J.-J. LETI/LIR's amorphous silicon uncooled microbolometer development. In Proceedings of the SPIE Aerospace/Defense Sensing and Controls, Orlando, FL, USA, 13–17 April 1998; Volume 3379.

64. Dong, L.; Yue, R.; Liu, L. Fabrication and characterization of integrated uncooled infrared sensor arrays using a-Si thin-film transistors as active elements. *J. Microelectromech. Syst.* **2005**, *14*, 1167–1177. [CrossRef]

65. Tissot, J.-L.; Tinnes, S.; Durand, A.; Minassian, C.; Robert, P.; Vilain, M.; Yon, J.-J. High-performance uncooled amorphous silicon video graphics array and extended graphics array infrared focal plane arrays with 17-µm pixel pitch. *Opt. Eng.* **2011**, *50*, 061006. [CrossRef]

66. Niklaus, F.; Decharat, A.; Jansson, C.; Stemme, G. Performance model for uncooled infrared bolometer arrays and performance predictions of bolometers operating at atmospheric pressure. *Infrared Phys. Technol.* **2008**, *51*, 168–177. [CrossRef]

67. Kruse, P.W. *Uncooled Thermal Imaging: Arrays, Systems, and Applications*; SPIE Press: Bellingham, WA, USA, 2001; Volume 51, p. 8.

68. Lloyd, J.M. *Thermal Imaging Systems*; Springer Science & Business Media: Berlin/Heidelberg, Germany, 2013.

69. Lee, H.; Yoon, J.; Yoon, E.; Ju, S.; Yong, Y.; Lee, W.; Kim, S. A high fill-factor IR bolometer using multi-level electrothermal structures. In Proceedings of the International Electron Devices Meeting, San Francisco, CA, USA, 6–9 December 1998; pp. 463–466.

70. Murphy, D.; Ray, M.; Wyles, R.; Asbrock, J.; Lum, N.; Wyles, J.; Hewitt, C.; Kennedy, A.; van Lue, D.; Anderson, J.; et al. High-sensitivity 25-micron microbolometer FPAs. In Proceedings of the SPIE AeroSense 2002, Orlando, FL, USA, 1–5 April 2002; Volume 4721.

71. Shen, N. Study on the Microbolometer in a Standard CMOS Technology. Ph.D. Thesis, Dalian University of Technology, Dalian, China, 2015.

72. Erturk, O.; Akin, T. Design and implementation of high fill-factor structures for low-cost CMOS microbolometers. In Proceedings of the 2018 IEEE Micro Electro Mechanical Systems (MEMS), Belfast, UK, 21–25 June 2018; pp. 692–695.

73. Tanrikulu, M.Y. Three-level microbolometer structures: Design and absorption optimization. *Opt. Eng.* **2013**, *52*, 083102. [CrossRef]

74. Schneeberger, N.; Paul, O.; Baltes, H. Spectral infrared absorption of CMOS thin film stacks. In Proceedings of the International Conference on Micro Electro Mechanical Systems, Orlando, FL, USA, 21 June 1999; pp. 106–111.

75. Luo, M. Research on Microbolometers Based on Standard Integrated Circuit Technology. Ph.D. Thesis, Nanjing University, Nanjing, China, 2020.

76. Ueno, M.; Kaneda, O.; Ishikawa, T.; Yamada, K.; Yamada, A.; Kimata, M.; Nunoshita, M. Monolithic uncooled infrared image sensor with 160 by 120 pixels. In Proceedings of the SPIE's 1995 International Symposium on Optical Science, Engineering, and Instrumentation, San Diego, CA, USA, 5 October 1995; Volume 2552.

77. Tanaka, A.; Suzuki, M.; Asahi, R.; Tabata, O.; Sugiyama, S. Infrared linear image sensor using a poly-Si pn junction diode array. *Infrared Phys.* **1992**, *33*, 229–236. [CrossRef]

78. Klaassen, E.H.; Reay, R.J.; Storment, C.; Kovacs, G.T.A. Micromachined thermally isolated circuits. *Sens. Actuators A Phys.* **1997**, *58*, 43–50. [CrossRef]

79. Rogalski, A.; Martyniuk, P.; Kopytko, M. Challenges of small-pixel infrared detectors: A review. *Rep. Prog. Phys.* **2016**, *79*, 046501. [CrossRef]

80. Zhang, X.; Liu, H.; Zhang, Z.; Wang, Q.; Zhu, S. Controlling thermal emission of phonon by magnetic metasurfaces. *Sci. Rep.* **2017**, *7*, 41858. [CrossRef]

81. Liu, T.; Qu, C.; Almasri, M.; Kinzel, E. Design and analysis of frequency-selective surface enabled microbolometers. In Proceedings of the SPIE Defense + Security, Baltimore, MD, USA, 17–21 April 2016; Volume 9819.

82. Luo, M.; Zhang, C.; Ren, F.; Yang, Q.; Lin, X.; Yan, F.; Ji, X. Metasurface-coupled microbolometers in CMOS technology. In Proceedings of the Applied Computational Electromagnetic Society Symposium, Beijing, China, 29 July–1 August 2018.

83. Abdullah, A.; Koppula, A.; Alkorjia, O.; Liu, T.; Kinzel, E.; Almasri, M. Metasurface integrated microbolometers. In Proceedings of the 2019 IEEE Research and Applications of Photonics in Defense Conference (RAPID), Miramar Beach, FL, USA, 19–21 August 2019.

84. Hanson, C.; Ajmera, S.; Brady, J.; Fagan, T.; McCardel, W.; Morgan, D.; Schimert, T.; Syllaios, A.J.; Taylor, M. Small Pixel a-Si/a-SiGe Bolometer Focal Plane Array Technology at L-3 Communications. In Proceedings of the SPIE Defense, Security, and Sensing, Orlando, FL, USA, 5–9 April 2010; Volume 7660.

85. Tanaka, A.; Matsumoto, S.; Tsukamoto, N.; Itoh, S.; Chiba, K.; Endoh, T.; Nakazato, A.; Okuyama, K.; Kumazawa, Y.; Hijikawa, M. Influence of bias-heating on a titanium bolometer infrared sensor. In Proceedings of the SPIE AeroSense'97, Orlando, FL, USA, 13 August 1997; Volume 3061, pp. 198–209.

86. Jo, Y.; Woo, D.H.; Lee, H.C. TEC-less ROIC with self-bias equalization for microbolometer FPA. *IEEE Sens. J.* **2015**, *15*, 82–88.

87. Eminoglu, S.; Gulden, M.A.; Bayhan, N.; Incedere, O.S.; Soyer, S.T.; Ustundag, C.M.; Isikhan, M.; Kocak, S.; Turan, O.; Yalcin, C.; et al. MT3825BA: A 384×288-25µm ROIC for Uncooled Microbolometer FPAs. In Proceedings of the SPIE Defense + Security, Baltimore, MD, USA, 5–9 May 2014; Volume 9070.

88. Tissot, J.-L.; Chatard, J.-P.; Fieque, B.; Legras, O. High-Performance and Low-Thermal Time Constant Amorphous Silicon-Based 320 × 240 Uncooled Microbolometer IRFPA. In Proceedings of the SPIE Photonics Asia, Beijing, China, 8 November 2004; Volume 5640.

89. Tissot, J.L.; Trouilleau, C.; Fieque, B.; Crastes, A.; Legras, O. Uncooled microbolometer detector: Recent developments at ULIS. *Opto Electron. Rev.* **2006**, *14*, 25–32. [CrossRef]

90. Mosavi, A.; Moezzi, M. A mismatch compensated readout IC for an uncooled microbolometer infrared FPA. *AEU Int. J. Electron. Commun.* **2020**, *123*, 153263. [CrossRef]

91. Chen, X.; Lv, Q. A versatile CMOS readout integrated circuit for microbolometric infrared focal plane arrays. *Optik* **2013**, *124*, 4639–4641. [CrossRef]

92. Tchagaspanian, M.; Villard, P.; Dupont, B.; Chammings, G.; Martin, J.; Pistre, C.; Lattard, D.; Chantre, C.; Arnaud, A.; Yon, J. Design of ADC in 25 µm pixels pitch dedicated for IRFPA image processing at LETI. In Proceedings of the SPIE Defense and Security Symposium, Orlando, FL, USA, 10–12 April 2007; Volume 6542.

93. Weiler, D.; Hochschulz, F.; Würfel, D.; Lerch, R.; Geruschke, T.; Wall, S.; Heß, J.; Wang, Q.; Vogt, H. Uncooled digital IRFPA-family with 17 µm pixel-pitch based on amorphous silicon with massively parallel Sigma-Delta-ADC readout. In Proceedings of the SPIE Defense + Security, Baltimore, MD, USA, 5 May 2014; Volume 9070.

94. Weiler, D.; Russ, M.; Würfel, D.; Lerch, R.; Yang, P.; Bauer, J.; Vogt, H. A digital 25 µm pixel-pitch uncooled amorphous silicon TEC-Less VGA IRFPA with massive parallel sigma-delta-ADC readout. In Proceedings of the SPIE Defense, Security, and Sensing, Orlando, FL, USA, 5–9 April 2010; Volume 7660.

95. Robert, P.; Dupont, B.; Pochic, D. Design trade-offs in ADC architectures dedicated to uncooled infrared focal plane arrays. In Proceedings of the SPIE Defense and Security Symposium, Orlando, FL, USA, 16–21 March 2008; Volume 6940.

96. Dupont, B.; Robert, P.; Dupret, A.; Villard, P.; Pochic, D. Model based on-chip 13bits ADC design dedicated to uncooled infrared focal plane arrays. In Proceedings of the SPIE Optics/Photonics in Security and Defence, Florence, Italy, 17–18 September 2007; Volume 6737.

97. Shafique, A.; Ceylan, Ö.; Yazici, M.; Kaynak, M.; Gurbuz, Y. A low-power CMOS readout IC with on-chip column-parallel SAR ADCs for microbolometer applications. In Proceedings of the SPIE. Defense + Security, Orlando, FL, USA, 15–19 April 2018; Volume 10624.

98. Scribner, D.A.; Kruer, M.R.; Killiany, J.M. Infrared focal plane array technology. *Proc. IEEE* **1991**, *79*, 66–85. [CrossRef]

99. Compact LWIR-Thermal Camera Core: Boson. Available online: www.flir.com (accessed on 20 July 2020).

100. Kimata, M.; Tokuda, T.; Tsuchinaga, A.; Matsumura, T.; Abe, H.; Tokashiki, N. Vacuum packaging technology for uncooled infrared sensor. *IEEJ Trans. Electr. Electron. Eng.* **2010**, *5*, 175–180. [CrossRef]

101. Li, C.; Han, C.J.; Skidmore, G.; Cook, G.; Kubala, K.; Bates, R.; Temple, D.; Lannon, J.; Hilton, A.; Glukh, K.; et al. Low-cost uncooled VOx infrared camera development. In Proceedings of the SPIE Defense, Security, and Sensing, Baltimore, MD, USA, 29 April–3 May 2013; Volume 8704.

102. Brady, J.; Schimert, T.; Ratcliff, D.; Gooch, R.; Ritchey, B.; McCardel, P.; Rachels, K.; Ropson, S.; Wand, M.; Weinstein, M.; et al. Advances in amorphous silicon uncooled IR systems. In Proceedings of the SPIE AeroSense'99, Orlando, FL, USA, 5–9 April 1999; Volume 3698.

103. Kennedy, A.; Masini, P.; Lamb, M.; Hamers, J.; Kocian, T.; Gordon, E.; Parrish, W.; Williams, R.; LeBeau, T. Advanced uncooled sensor product development. In Proceedings of the SPIE Defense + Security, Baltimore, MD, USA, 20–24 April 2015; Volume 9451.

104. Dumont, G.; Rabaud, W.; Yon, J.-J.; Carle, L.; Goudon, V.; Vialle, C.; Becker, S.; Hamelin, A.; Arnaud, A. Current progress on pixel level packaging for uncooled IRFPA. In Proceedings of the SPIE Defense, Security, and Sensing, Baltimore, MD, USA, 25–26 April 2012; Volume 8353.

105. Yon, J.; Dumont, G.; Goudon, V.; Becker, S.; Arnaud, A.; Cortial, S.; Tisse, C. Latest improvements in microbolometer thin film packaging: Paving the way for low-cost consumer applications. In Proceedings of the SPIE Defense + Security, Baltimore, MD, USA, 5–9 May 2014; Volume 9084.

106. Dumont, G.; Arnaud, A.; Imperinetti, P.; Mottin, E.; Simoens, F.; Vialle, C.; Rabaud, W.; Grand, G.; Baclet, N. Innovative on-chip packaging applied to uncooled IRFPA. In Proceedings of the SPIE Photonics Asia 2007, Beijing, China, 11–15 November 2007; Volume 6835.

MDPI

St. Alban-Anlage 66

4052 Basel

Switzerland

Tel. +41 61 683 77 34

Fax +41 61 302 89 18

www.mdpi.com

Micromachines Editorial Office

E-mail: micromachines@mdpi.com

www.mdpi.com/journal/micromachines